At Issue

Should the Government
Fund Embryonic
Stem Cell Research?

Other Books in the At Issue Series:

At Issue

Should the Government Fund Embryonic Stem Cell Research?

Amy Francis, Book Editor

GREENHAVEN PRESS
A part of Gale, Cengage Learning

GALE
CENGAGE Learning

Detroit • New York • San Francisco • New Haven, Conn • Waterville, Maine • London

Christine Nasso, *Publisher*
Elizabeth Des Chenes, *Managing Editor*

© 2009 Greenhaven Press, a part of Gale, Cengage Learning.

Gale and Greenhaven Press are registered trademarks used herein under license.

For more information, contact:
Greenhaven Press
27500 Drake Rd.
Farmington Hills, MI 48331-3535
Or you can visit our Internet site at gale.cengage.com

For product information and technology assistance, contact us at

Gale Customer Support, 1-800-877-4253
For permission to use material from this text or product, submit all requests online at
www.cengage.com/permissions

Further permissions questions can be emailed to permissionrequest@cengage.com

Articles in Greenhaven Press anthologies are often edited for length to meet page requirements. In addition, original titles of these works are changed to clearly present the main thesis and to explicitly indicate the author's opinion. Every effort is made to ensure that Greenhaven Press accurately reflects the original intent of the authors. Every effort has been made to trace the owners of copyrighted material.

Cover image ©Images.com/Corbis.

LIBRARY OF CONGRESS CATALOGING-IN-PUBLICATION DATA

Should the government fund embryonic stem cell research? / Amy Francis, editor.
  p. cm. -- (At issue)
  Includes bibliographical references and index.
  ISBN-13: 978-0-7377-4438-5 (hardcover)
  ISBN-13: 978-0-7377-4439-2 (pbk.)
  1. Embryonic stem cells--Research--United States--Finance. 2. Federal aid to medical research--United States. I. Francis, Amy.
  QH588.S83S56 2009
  174.2--dc22

                                               2009012035

Printed in the United States of America
1 2 3 4 5 6 7 13 12 11 10 09

Contents

Introduction

Stem cell researchers were using stem cells to find cures long before this research caught the attention of the mass media or became a topic of controversy. At least 40 diseases and conditions, including juvenile diabetes and some cancers, are commonly treated using adult stem cells, and trials for many new treatments are constantly underway.

Although treatments using stem cells from bone marrow have been used for years, scientists are still forging new ground with these cells. As reported on January 29, 2009, in the British daily newspaper, *Telegraph*, a clinical trial showed multiple sclerosis symptoms could be reversed using a patient's own bone marrow stem cells. The study, led by Dr. Richard Burt from the Northwestern University Feinberg School of Medicine in Chicago, reported that the procedure "not only seems to prevent neurological progression, but also appears to reverse neurological disability." Of the 21 patients who were given this stem cell treatment, 17 cited improvement in their symptoms, which no previous treatment was able to provide.

Researchers are making progress using stem cells from other areas of the body, and they are even finding some cells may have the ability to form tissue other than their source tissue. For example, cord blood, blood from the umbilical cord left over after birth, is not only rich in stem cells but has the advantage of less stringent matching criteria between donor and recipient than bone marrow stem cells. This type of research is leading to medical breakthroughs never before thought possible.

As reported in the British medical journal, *The Lancet*, on June 9, 2007, researchers found that umbilical cord blood is as successful as bone marrow transplants in treating acute childhood leukemia. Further studies are showing promise with this treatment for leukemia in adults as well.

Cord blood is also showing promise in other areas. In 2008, Maia Friedlander became the first child from New Zealand to be treated with her own cord blood. Only twelve weeks after receiving a transfusion of her stem cells from cord blood, her father was quoted in *Scoop*, a New Zealand news source, stating, "Just a few days later after the procedure her eyes started to look more alert, and she lost the unfocused, dreamy kind of look she had always had. Her arms and legs began to straighten out, and her physical coordination improved."

Stem cell research is also helping scientists make progress in the area of organ and tissue transplants. On December 20, 2008, *The Lancet* highlighted a successful trachea transplant— the first tissue transplant to use stem cells: "On a scientific level, the procedure has proved that it is possible to grow organs using a patient's own stem cells, eliminating the problems of rejection that have always plagued transplants. Stem cell research, which has promised so much in the laboratory, has at last delivered a genuine clinical advance."

More clinical trials are starting every day. According to its Web site, the University of Utah is performing the first trial in the United States to treat heart failure using a patient's own stem cells. As reported in the British *Telegraph*, research being done at Imperial College in London to rebuild heart tissue damaged during a heart attack could end the need for transplants: "They have discovered a way to extract, grow in the laboratory and then graft on a patient's own muscle-building cells which can be used to patch up the heart and increase its pumping power." Trials on humans could be as little as two years away.

Although therapies exist for a number of conditions, severe injuries to some areas of the body, such as the spinal cord, which can leave patients paralyzed, still have no existing curative treatment either through stem cells or other methods.

The potential of embryonic stem cells to be used directly in treatment or used for research to help scientists better understand adult stem cells has led to significant controversy. Both the promise and controversy surrounding embryonic stem cell research, particularly as they relate to public funding, are explored by the authors in *At Issue: Should the Government Fund Embryonic Stem Cell Research?*

1

The United States Cannot Remain Neutral on Stem Cell Research

President's Council on Bioethics

The President's Council on Bioethics *is a panel of experts whose purpose is to advise the president of the United States on ethical issues related to advances in biomedical science and technology.*

As the federal government continues to revisit the current regulations impacting the funding of stem cell research, it needs to be wary of the arguments of both the opponents and proponents of the issue. While one side views embryonic stem cell research as murder and the other as engaging in life-saving research, neither side adequately addresses the key issue of funding, which does not include regulations on stem cell research done with private money. Since funding an activity is one way the government can give greater significance to it, supporting stem cell research with federal funds is not something the government should do lightly. Further, the U.S. government cannot hold a policy of neutrality with regard to embryonic stem cell research, as the public tends to view this policy as siding with those who oppose such research.

The national debate over embryonic stem cell policy often raises the most fundamental questions about the status of the human embryo and the legitimacy of research that de-

"The Administration's Human Stem Cell Research Funding Policy: Moral and Political Foundations," *President's Council on Bioethics*, September 2003. Reproduced by permission.

stroys such embryos. For those caught up in this debate, it is easy to forget that the question at issue is not whether research with embryos should be allowed, but rather whether it should be financed with federal taxpayer dollars.

Defining the Debate

The difference between prohibiting embryo research and refraining from funding it has often been intentionally blurred by both sides to the debate. Proponents of funding want as much material support as possible for research efforts, and so they play down the availability of other sources of funding, so as to make the case that the work could not proceed without public money. They wish to portray their adversaries as opposed to important scientific progress, and so it is useful to them to shift the grounds of the debate in the direction of an argument about the legitimacy of the research itself, rather than the meaning of paying for it with taxpayer funds. All too often, they claim that current policy in fact prohibits all stem cell research. Meanwhile, most opponents of funding for embryo research are, in fact, also opponents of embryo research more generally, and they use the debate over funding as a forum for making their case for the moral status of the embryo. Some of them would be only too glad to see the research banned. The question of funding itself is therefore rarely taken up in full.

The Role of the Government in Scientific Research

That question arises because modern governments do more than legislate and enforce prohibitions and limits. In the age of the welfare state, the government, besides being an enforcer of laws and a keeper of order, is also a gargantuan provider of resources. Political questions today, therefore, reach beyond what ought and ought not be allowed, to include questions of what ought and ought not be encouraged, supported, and made possible by taxpayer funding. The decision to fund an activity is more than an offer of resources. It is also a declaration of official national support and endorsement, a positive assertion that the activity in question is deemed by the nation as a whole, through its government, to be good and worthy. When something is done with public funding, it is done, so to speak, in the name of the country, with its blessing and encouragement.

The American taxpayer is by far the greatest benefactor of science in the world.

To offer such encouragement and support is therefore no small matter. Aside from its material importance, the offer is also laden with moral and political meaning. In the age of government funding, the political system is sometimes called upon to decide not only the minimal standards of conduct, but also the maximum standards of legitimacy and importance. When the nation decides an activity is worth its public money, it declares that the activity is valued, desired, and favored.

The United States has long held the scientific enterprise in such high regard. Since the middle of the twentieth century, the federal government, with the strong support of the American people, has funded scientific research to the tune of many hundreds of billions of dollars. The American taxpayer is by far the greatest benefactor of science in the world. This is so because the American public greatly values the contributions of science to human knowledge, human power, human health, and the standard of living. We Americans have always been boosters of science and medicine, deeming it worthy of support for moral as well as material reasons.

No Middle Ground

But this enthusiasm for science is not without its limits. As already noted, we attach restrictions on federally funded research, for instance to protect human subjects. We also put limits on some practices that might offer life-saving benefits, for example by prohibiting the buying and selling of organs for transplantation. Also, as in the present case, many Americans have specific moral reasons for opposing certain lines of research or clinical practice, for example those that create human life in the laboratory or that involve the exploitation and destruction of human fetuses and embryos. The two sides of the embryo research debate tend to differ sharply on the fundamental moral significance of the activity in question. One side believes that what is involved is morally obnoxious in the extreme, and indeed may be akin to homicide; while the other believes embryo research is noble or even heroic, and is worthy of praise and support. It would be very difficult for the government to find a middle ground between these two positions, since the two sides differ not only on what should or should not be done, but also on the attitude from which the activity should be approached.

The present policy is an articulation of the view that public resources will not be used to encourage the destruction of human embryos.

Remaining Neutral

To this point, the only workable approach found has been the policy of federal neutrality, whereby the federal government does not prohibit embryo research, but also does not officially condone it, encourage it, or support it with public funds.[1] This has allowed the political system to avoid forcing the

1. On March 9, 2009, President Barack Obama reversed the ban on federally funded embryonic stem cell research in an executive order calling for the National Institutes of Health to create guidelines for funding "responsible, scientifically worthy" human embryonic stem cell research.

question of whether embryo research is good or evil, to avoid compelling those citizens who oppose it to fund it with their tax money, and to avoid banning it against the wishes of those who believe it serves an important purpose. The approach is based, at least in part, on the conviction that debates over the federal budget are not the place to take up the anguished question of the embryo.

But the position is not only a compromise between those who would have the government bless and those who would have the government curse this activity. It is also a statement of a certain principle: namely, that public sanction makes a difference. The present policy is an articulation of the view that public resources will not be used to encourage the destruction of human embryos. While embryo destruction may be something that some Americans support and engage in, it is not something that America as a nation has officially supported or engaged in. It has generally not been deemed to meet the standard for public acclaim and taxpayer funding.

Of course, if the funding issue were merely a proxy for the larger dispute over the moral status of the embryo, then the principle of federal neutrality would appeal only to those who would protect the embryo, and would succeed only as long as they were able to enact it. The argument might end there, with a vote-count on the question of the moral status of the embryo. But some proponents of the present law suggest that the particulars and contours of the embryo research debate offer an additional rationale for the principle of federal neutrality on embryo research. Here again, it is important to remember that the issue in question is public funding, not permissibility. Opponents of embryo research have in most cases acquiesced in narrowing the debate to the question of funding. They do not argue for a wholesale prohibition of embryo research by national legislation, though many of them see such work as an abomination and even a species of homicide.

In return, proponents of the Dickey Amendment[2] argue that it would be appropriate for supporters of research to agree to do without federal funding. Many Americans believe that this is a life and death question, and do not want their country, in their name, to take the wrong side. Even for those who disagree, the character and intensity of feeling among fellow citizens on the other side surely must count for something.

Considering Funding in the Future

On the other hand, it might reasonably be argued that part of living under majority rule is living with the consequences of sometimes being in the minority. Were the Congress to overturn the current policy of federal neutrality, opponents of funding for embryo research would not be alone in being compelled to pay for activities they abhor. We all see our government do things, in our name, with which we disagree. Some of these might even involve life and death questions, for instance in wars that some citizens oppose. The existence of strong moral opposition to some policy is not in itself a decisive argument against proceeding with that policy. But once more it is worth reminding ourselves that what is at issue is public funding of a practice that could be (and is) otherwise privately funded. The fact that another source of funding exists might reasonably set the bar somewhat higher for involving the public in the encouragement and facilitation of a highly controversial practice. If so, proponents of embryo research might need to show not only that it is a legitimate practice, but also that they are unable to procure sufficient funds from private sources, that the practice is so important to the public good that it should override the strong moral

2. The Dickey Amendment, a rider to an appropriations bill for the Department of Health and Human Services (HHS), was passed in 1995. It prevents the federal funding of embryonic stem cell research, and it has been renewed with each new HHS appropriation.

objections of a very substantial portion of the public, and that no morally less problematic alternatives exist to accomplish the same ends.

These concerns and arguments give the question of funding its own crucial significance, even apart from the more fundamental question of the legitimacy and propriety of the act being funded.

2

The Majority of Americans Want Stem Cell Research to Be Federally Funded

Pam Solo and Gail Pressberg

Pam Solo is founder and president of the Civil Society Institute and author of From Protest to Policy: Beyond the Freeze to Common Security. *Gail Pressberg is a senior fellow at the Civil Society Institute and works on the institute's stem cell advocacy program.*

Although the general public continues to voice support for stem cell research, there has been little support from the United States government for such research. Further, the United States now lags behind other nations that support stem cell research. For this reason, public advocacy groups and state government officials are working to help fund research at the local level and ensure progress is not lost without federal funding. California's successful passage of the Stem Cell Research and Cures Bond Act of 2004 brought not only additional funding and jobs to the state, but also the possibility of more progress toward stem cell therapies. While grassroots groups attempt to sway officials to take action, other states are attempting to pass similar types of bonds. Continuing the debate on stem cell research is needed to educate citizens and will result in the best plan to support this emerging science.

Pam Solo and Gail Pressberg, "Will Patients and Their Allies Prevail?" *The Promise and Politics of Stem Cell Research*. Westport, CT: Praeger, 2007. Copyright © 2007 by Pam Solo and Gail Pressburg. All rights reserved. Reproduced by permission of Greenwood Publishing Group, Inc., Westport, CT.

The future of stem cell research in the United States will be determined by two powerful social forces: the market and pressure from patients and their advocates. As stem cell research advances, the interest of investors grows. Major pharmaceutical companies are joining the smaller biotech companies in investing in the research. Venture capitalist Steven Burrill estimates that as many as a hundred companies in the United States and an equal number elsewhere are engaged in stem cell research.

The Fate of Stem Cell Research in the U.S.

California's successful Proposition 71 [Stem Cell Research and Cures Bond Act of 2004] has had a profound influence on the market. Although a lawsuit has prevented the state from issuing any stem cell research grants thus far, California's favorable stem cell policies have attracted new investment. Advanced Cell Technology, a biotech company from Massachusetts, has opened an office in northern California. This has brought the state new jobs and the potential of new discoveries. Philanthropists Eli and Edythe Broad have donated $25 million to develop a new stem cell center at the University of Southern California. This center hopes to collaborate with nearby research facilities to create a stem cell research hub.

Not wanting to be left behind, other states have been seeking ways to provide public funds for stem cell research. Internationally, congenial policy environments in several countries are setting the stage for progress, whether the United States participates or not. The leading industrialized country in the world could find itself marginalized in what promises to be an economic boom at least as big as that sparked by the semiconductor and the microchip. That prospect may well spur the United States to change its policies.

The second major factor in deciding the fate of stem cell research will be patients and the grassroots organizing that will eventually push elected officials to take decisive action. The public's loyalty is not to the Democrats or the Republicans, but to the ideal of government action that is appropriate and proportional to the problems at hand.

Survey after survey shows that Americans are not nearly as divided on the major issues as the news media or politicians paint them to be.

Public Opinion and Politics

Washington has recently been dominated by politicians who are out of step with what a majority of Americans believe about embryonic stem cell research as well as many other issues. The political class is engaged in one conversation, while the rest of the country is having an altogether different conversation. A massive chasm exists between the elites—that is, the politicians, people who work to elect them, and the lobbyists who are hired to influence them—and ordinary American citizens, who are interested in solving problems that affect their lives. Survey after survey shows that Americans are not nearly as divided on the major issues as the news media or politicians paint them to be. This is true with regard to abortion, health care reform, stem cell research, and environmental issues, to name a few. Americans want leadership, and they want action.

Nevertheless, policies can change in response to the market or to the overwhelming demands of the public. The market may well eclipse the political process, but the political process is precisely what the age of regenerative medicine needs. Science needs society as much as society needs science.

The Competitive Field

The antiscience policies of [former president] George W. Bush's administration threaten to undermine American competitiveness in a number of fields beyond stem cell research.

Global research-and-development budgets from our major trading competitors emphasize science and technology in dealing with environmental challenges, global warming, and alternative and renewable energy sources. Alan Leshner, the CEO of the American Association for the Advancement of Science, recently said, "My friends in the European Union [E.U.] are ecstatic. . . . Between the visa problems—preventing the best students from China and India from entering this country to study or work—and the stem cell ban, our competitors are just in heaven. While we are restricting research, the E.U. is working hard on ways to increase mobility. . . . They are not fools." Once again, the market may lead the way, while politics and policy lag behind knowledge, analysis, and new technologies. For example, [investment banks] J. P. Morgan and Goldman Sachs have now determined that global warming entails an economic risk to investors and shareholders that must be calculated in investment and lending strategies and decisions.

More than 300 patients' organizations, medical schools, and scientific organizations . . . demanded that the budget allocation for NIH be raised.

Changing the Political Field

Politicians are supposed to lead. They run for office and are elected to provide leadership on major problems and find practical steps toward their solutions. Instead, the permanent campaign, the constant fundraising, and the posturing for media attention occupy much of an elected official's time and attention. Politicians tend to make policy by listening to polls, major donors, and lobbyists, instead of considering what is best for the country. The only antidote to these dynamics is for citizens to let the politicians know that the price they will pay for bad policies is to be voted out of office.

Advocacy from the Ad Hoc Group

The federal budget debate in 2006 drew a strong response from patients, scientists, and the medical community. Facing a soaring budget deficit because of tax cuts and the war in Iraq, President [George W.] Bush froze funding for the National Institutes of Health [NIH], the government body that makes grants for medical research. (NIH's budget has steadily decreased since fiscal year 2003.) Enter patients and their allies in academia and Congress. More than 300 patients' organizations, medical schools, and scientific organizations formed the Ad Hoc Group for Medical Research, which demanded that the budget allocation for NIH be raised at least enough to account for the biomedical inflation index, estimated to be 3.2 percent. Meanwhile, Republican Senator Arlen Specter and Democratic Senator Tom Harkin introduced an amendment requesting an additional $7 billion for medical research and health purposes. Although this amendment was not the actual appropriation for medical research in the budget process, it limits how much can later be appropriated.

The member organizations of the Ad Hoc Group sent out action alerts to individuals across the United States, urging patients and scientists to call on senators to pass the Specter-Harkin amendment. On March 16, 2006, the amendment passed the Senate by a vote of seventy-three to twenty-seven. Ellen Stovall, a 34-year cancer survivor and the president and CEO of the National Coalition for Cancer Survivorship, summed up the victory in an e-mail to her membership: "By adopting this amendment the Senate has restored room in next year's budget for maintaining funding for crucial cancer research programs. This is an important first step toward retaining this important funding. The appropriations process in Congress is complex and many steps lay ahead. We will keep you informed about the status of research funding and if it is threatened again in the coming months we will ask you to speak up again."

The Stem Cell Research Enhancement Act

The Coalition for the Advancement of Medical Research and many coalitions of patients across the United States succeeded to pass the Stem Cell Research Enhancement Act in both the House of Representatives and the Senate. After Congress passed the Stem Cell Research Enhancement Act California Governor Arnold Schwarzenegger urged President [George W.] Bush not to veto the bill. He wrote to President Bush, "Mr. President, I urge you not to make the first veto of your presidency one that turns America backwards on the path of scientific progress and limits the promise of medical miracles for generations to come." Unfortunately President Bush did not heed the Governor's advice.

When science is attacked on purely ideological grounds, its very integrity is at risk.

Support from the States

When President Bush vetoed the Stem Cell Research Enhancement Act, governors of several states, including Republicans and Democrats, responded by allocating state funds for stem cell research. Governor Arnold Schwarzenegger of California committed $100 million from the state's general funds to pay for stem cell research. Governor Rod Blagojevich, a Democrat, immediately allocated $5 million for stem cell grants in Illinois. Candace Coffee, a Los Angeles resident quoted in the *New York Times* on July 25, 2006, who suffers from Devic's disease [an autoimmune disease], summed up the sentiments of most patients when she appeared at a press conference in Los Angeles with Governor Schwarzenegger. Ms. Coffee stated to the *New York Times* that "President Bush's veto stole my hope. But just as quickly as our hope was stolen, it was renewed." At the state level, patients' organizations have also succeeded. In Missouri, the Coalition for Lifesaving Cures won a case in court against a right-wing organization that plagiarized its Web site in an attempt to confuse voters about the stem cell initiative on the ballot for November 2006. In Illinois, patient activists got the state's funding for stem cell research increased from $10 million in 2005 to $100 million for 2007. Patient activists in Maryland got the state legislature to approve funding for stem cell research, and New Jersey increased its allocation for stem cell research.

Additional Threats to Science

Despite the successful campaigns of patients' groups and their allies, threats to cell biology still exist. A near majority of United States senators would make somatic cell nuclear transfer—cloning solely for medical-research purposes—a federal crime. Professor Paul Berg, a 1980 Nobel Laureate in chemistry for his work on recombinant DNA, summed up the challenges that the patients' community, their scientific allies, and the public have in campaigning against new developments in biomedical research: "The threat is not only to stem cell biologists. It affects the entire scientific community. For when science is attacked on purely ideological grounds, its very integrity is at risk. . . . Such a challenge to basic scientific research would have been unthinkable a few decades ago. Following World War II, the federal government enthusiastically embraced untargeted research, what some often refer to as curiosity-driven research. The achievements of our basic research enterprise made us the world leader in the physical and life sciences, fueling the economic prosperity that the country enjoyed for the last half of the 20th century. . . . The respect awarded scientists could still be seen in the middle and late 1970s when for the first time federal officials considered prohibiting a line of basic research—the study of recombinant DNA technology. Although a number of lawmakers feared that unfettered pursuit of 'genetic engineering' might result in unforeseen and damaging consequences for human health and the Earth's ecosystems, proposals to restrict such work were stymied by scientists' opposition. Instead, largely at the urging and guidance of the scientific community, research was permitted to proceed according to federal guidelines that mandated oversight by an institutional review process and approval by the National Institutes of Health.

The debate about stem cell research, in particular, and regenerative medicine generally must continue in the most informed and public manner possible.

Why the Stem Cell Debate Is Unique

"Today, there is hardly a field in the life sciences that has not been transformed by our enhanced understanding of DNA. . . .

"Unlike the debate over the recombinant DNA issue [whose opponents believe it to be an environmental threat] . . . the concerns about embryonic stem cell research have more to do with values and ideology. The essence of the confrontation is that the only available source of human embryonic stem cells is the very early stage embryo—the blastocyst, a cell no larger than a period mark in this newspaper. Consequently, this research challenges some people's deeply held religious views about the moral status of the early embryo.

"What is so troubling is that the quality of the science being planned may no longer be the principal determinant in whether that line of research should be permitted. Rather, theological and ideological pronouncements are increasingly taking over. Social conservatives and their political representatives are poised to define the boundaries and even the permissibility of scientific research. . . . We hear calls to use the power of the public purse to limit certain lines of inquiry, as now exists for human embryo research."

The debates about the relationships between science and society will only intensify in the decades to come, and we applaud this. The debate about stem cell research, in particular, and regenerative medicine generally must continue in the most informed and public manner possible. Only with an engaged and fully educated citizenry can we arrive at policies that will encourage the best science has to offer and navigate through the complex moral, legal, and ethical issues the science presents.

3

Americans Are Misinformed About Stem Cell Research

Michael Fumento

Michael Fumento is an attorney, photographer, and journalist based in Arlington, Virginia, who specializes in health and science issues. He is the author of five books.

While embryonic stem cells may carry the potential to treat diseases, therapeutic applications are still several years out. Meanwhile, adult and other non-embryonic stem cells are already being used to treat, and in some cases cure, numerous diseases with additional applications currently being evaluated in clinical trials with humans. Despite the fact that more progress is being made with non-embryonic stem cells than embryonic stem cells, funding efforts continue to focus on embryonic stem cell research. The media is also choosing to focus on the potential of embryonic stem cells rather than the proven treatments available with non-embryonic stem cells. Funding dollars should be aimed at non-embryonic stem cell research, where more treatments and cures are within reach.

While the Democratic-controlled House [of Representatives] voted [January 2007] 253-174 to expand federal funding for embryonic stem-cell research, it fell far short of the 290 votes needed to override a virtually guaranteed presidential veto. A tragedy for victims of everything from

Alzheimer's to warts? Not at all. Each year there are stunning breakthroughs with adult stem cells, and 2007 has already brought its first.

Adult Stem Cells

Adult stem cells cure and treat more than 70 diseases and are involved in almost 1,300 human clinical trials. Scientists also keep discovering that adult stem cells are capable of creating a wider variety of mature cells. Perhaps the most promising of these was announced in the January issue of [scientific journal] *Nature Biotechnology*.

Anthony Atala, director of the Institute for Regenerative Medicine at Wake Forest University School of Medicine [Winston-Salem, North Carolina], reported that stem cells in the amniotic fluid that fills the sac surrounding the fetus may be just as versatile as embryonic stem cells. At the same time they maintain all the advantages that have made adult stem cells such a success.

That leaves embryonic stem cells with only one possible advantage—potential.

This has caused great consternation on the part of those seeking increased taxpayer embryonic stem cell funds. The reason is that there are currently no practical applications for this type of cell. There hasn't even been a single clinical trial revolving them. Researchers admit we won't have approved embryonic stem cell treatments for at least 10 years.

One advantage of embryonic stem cells has been that most types of adult stem cells cannot be multiplied outside of the body for very long, while embryonic ones may replicate in the lab indefinitely. But Atala's new amniotic stem cells grow as fast outside the body as embryonic stem cells (doubling every 36 hours), and he's now been growing the same cell line for two years, with no indication of slowing.

Embryonic Stem Cell Misinformation

That leaves embryonic stem cells with only one possible advantage—potential. Embryonic stem cells can be "differentiated" into all three "germ layers," or subtypes of cell. That means they should be able to be made into all of the 220 types of cells in humans. For a long while, adult stem cells were believed to be only capable of differentiation to a limited number of mature cells, depending on the type of adult stem cell with which you start. For example, a marrow cell could become any number of types of marrow or blood cells, but it couldn't become a muscle cell. That's a different germ layer.

Yet it's been virtually a state secret that for over five years researchers, beginning with a team headed by physician Catherine Verfaillie of the University of Minnesota Stem Cell Institute, have been reporting numerous types of adult stem cells (she used those from marrow) that in the lab could form mature cells from three germ layers. Experiments around the world have clearly shown that adult stem cells from one germ layer can be converted into those of another in a living human, such as those that turned marrow cells into heart muscle and blood vessels in live humans.

Amniotic Stem Cells

That said, amniotic stem cells may be the most easily differentiated of all—as well as among the easiest to extract in large amounts. Indeed, they are routinely recovered with a hypodermic needle during amniocentesis [a common medical procedure performed on pregnant women]. While it's widely believed that this procedure slightly increases the chance of miscarriage, a sizable study last November of 35,000 women who underwent mid-trimester testing found "no significant difference in loss rates between those undergoing amniocentesis and those not undergoing amniocentesis."

There are over four million births each year in the United States, yet Atala calculates that merely 100,000 amniotic stem cell specimens could supply 99 percent of the U.S. population's needs for perfect matches for transplants. (That assumes a perfect match is even needed.) About 700,000 amniocentesis procedures are performed in the United States and Western Europe each year.

Stem Cells in the Media

Some embryonic stem cell researchers have downplayed the Atala findings. The work will "still require a lot of replication from other groups before they can be conclusive," Stephen Minger, an embryonic stem cell scientist identified only as a "lecturer in stem-cell biology" told a British newspaper. "They have only shown that these particular stem cells can turn into a couple of different types of other stem cells. I would say that a hell of a lot more work is required." Other media outlets would say the same. *Newsweek International* claimed, "Many scientists are quick to emphasize that comprehensive human trials are still many years away."

Scientifically, all embryonic stem cells tend to become cancerous.

The *New York Times* refused even to allow people to read between the lines—they simply never reported the news about Atala's work. When a reader complained to the "Public Editor," an online ombudsman, about the omission, the *Times* responded that its genetics reporter, Nicholas Wade, "looked at the Atala paper last week and deemed it a minor development." Wade said of the paper, "It reports finding 'multipotent' stem cells in amniotic fluid. Multipotent means they can't do as much as bona fide embryonic stem cells (which are called 'pluripotent')."

Neither Minger nor *Newsweek* nor Wade could be more wrong. As Atala told PBS's *Online NewsHour*, "We have been able to drive the cell to what we call all three germ layers, which basically means all three major classes of tissues available in the body, from which all cells come from." I pointed out in a response to the *New York Times* posting that merely reading the online abstract of the Atala paper indicated the same. Of course, this is the same [news]paper that told readers in 2004 that there were no cures or treatments with adult stem cells. Not 70 cures or treatments, some dating back half a century—none.

It is neither paranoia nor exaggeration to say that the *New York Times* is engaged in a stem-cell cover-up.

Non-embryonic stem cell researchers are already performing miracles, such as growing new heart and liver tissue and treating multiple sclerosis—all in living humans.

Adult vs. Embryonic Stem Cells

What makes all of this worse is that Atala's work actually is a replication of numerous studies. He's just taken the research further and pulled his cells from amniotic fluid, whereas others have pulled the identical cells from the placenta. Amniotic and placenta stem cells are the same, as Atala himself noted. And as to human trials being "many years away," *Newsweek* is correct only if "years away" means "years ago." The *New England Journal of Medicine* carried one paper on a placenta stem cell trial back in 1996 and another paper two years later. There's been one ongoing clinical trial since 2001 to treat sickle cell anemia. The *Washington Post's* Rick Weiss, who has been accused of boosterism for embryonic stem cell research, tried to find a middle ground, saying that "The new [sic] cells are adding credence to an emerging consensus among experts that the popular distinction between embryonic and 'adult' stem cells—those isolated from adult bone marrow and other organs—is artificial."

Actually, what's "artificial" is the term "adult stem cell," which worked fine not so long ago when all adult stem cells were all pulled from bone marrow, but is confusing now that they're being extracted from placentas, amniotic fluid, and umbilical cords, which aren't exactly "adult" sources. But for discussions both scientific and moral, stem cells can still be broken down between the embryonic and the non-embryonic.

Problems with Embryonic Stem Cells

Scientifically, all embryonic stem cells tend to become cancerous; they require permanent, dangerous, immunosuppressive drugs because the body rejects them as foreign; and they are difficult to differentiate into the needed type of mature cells. Non-embryonic stem cells, however, do not become cancerous; they are far less likely to cause rejection (especially the youngest, including umbilical cord and amniotic/placenta); and they have been used therapeutically since the late 1950s (originally for leukemia) because they have the amazing ability to form the right type of mature cell merely upon being injected into a body that needs that type of cell.

It is these biological differences that have held embryonic stem cell research back, not a lack of federal funds.

As stem-cell researcher Malcolm Alison of the University of London told the *Daily Mail*, the amniotic cells "appear to be at least as malleable as embryonic stem cells but without all the ethical baggage."

Success with Non-embryonic Stem Cells

For all the talk over the morality of using human embryos in medicine, perhaps there's another moral issue at play: Non-embryonic stem cell researchers are already performing miracles, such as growing new heart and liver tissue and treating multiple sclerosis—all in living humans. Yet they struggle to get federal funds for their research. Given the growing number of state initiatives that fund embryonic stem cell, but not non-embryonic stem cell, research and given that overall National Institutes of Health funding increases are unlikely anytime soon, is it truly moral to take away funds from a technology that's been saving lives for half a century in favor of another technology that promises nothing but "promise"?

4

President Bush Should Not Have Vetoed Stem Cell Research Funding

Joseph J. Seneca

Joseph J. Seneca is chairman of the New Jersey Council of Economic Advisors and a professor at the Edward J. Bloustein School of Planning and Public Policy. He is the author of numerous works on state economic development and policy.

President George W. Bush's veto in 2006 of a bill that would have led to more federal funding for stem cell research was a major setback. In spite of this, states can minimize the impact the lack of funding, scientific support, and ethical oversight has had on research. Some states have already made progress with funds made available at the state level. Others, like New Jersey, have provided funding for stem cell research but spending was stalled in disputes as they attempted to launch new programs and build new facilities. New Jersey needs to take immediate steps to move forward, and other states looking to fund stem cell research in lieu of federal funding programs must also be careful to avoid such pitfalls.

It is incomprehensible that in the 21st century, the government of the United States is overtly restricting the advancement of a science that holds such promise to alleviate human pain and suffering. President [George W.] Bush's recent [2006] veto, the very first of his presidency, of legislation to increase federal funding for new lines of embryonic stem cell research is a major disappointment.

It is also a major inconsistency, since federal funding for such research continues to be provided for an original group of cell lines created before August 2001. What this country has just lost is the opportunity to provide substantial federal funding for research on additional embryonic stem cell lines that have greater possibilities than the original lines to yield effective therapies for a wide range of devastating human diseases and injuries.

One danger posed by the lack of enhanced federal support is that the stem cell initiatives in various states may fall victim to cyclical state budget constraints.

Stem Cell Research Continues Despite Veto

Stem cell research will still go forward in the United States, but it will do so, for now, without the supportive environment, coordinated scientific peer review process and ethical oversight of the National Institutes of Health. A number of states, along with private sources of funding, will continue to fill the vacuum created by this oxymoron [inherent contradiction] of a faith-based federal science policy. It is inevitable that, in time, the federal government will change this restriction. The tragedy is the lost opportunities and continued deep suffering of the victims and their families, absent the hope that such additional, federally supported, research would provide.

New Jersey has been a leader in all forms of stem cell research. It was the first state, under [former] Gov. Richard J. Codey, to provide research funds through a careful and impeccable ethical and scientific review process overseen by the New Jersey Commission on Science and Technology.

But pending state legislation to build and equip a stem cell institute and provide substantial research funding has languished for several years in the New Jersey Legislature in an unseemly dispute over the number of centers, their location and the allocation of funds.

What States Can and Should Do

A significant number of other states have committed resources to stem cell research and a competitive, though uncoordinated, race among these states is now well under way. California has accelerated its substantial commitment, given the recent veto by President Bush. Other countries, unconstrained by the self-imposed obstacles created in the United States, continue to devote significant resources to developing stem cell therapies. Researchers in these states and countries will have access to the new cell lines and will benefit from knowledge not available to those who must, because of the availability of existing federal funding, continue to work only with the original stem cell lines. What should be New Jersey's approach in this unnecessarily confused and cluttered situation? First, New Jersey, a state with a long legacy of scientific innovation and discovery, should pursue stem cell research aggressively and promptly. Building a first-class research facility alongside available clinical opportunities is the key priority that can wait no longer.

Economic Benefits of Stem Cell Research

New Jersey's economy would benefit at a time when several of its former leading technology industries are under severe competitive pressure. New science-based industries need to be created to replace other industries inevitably diminished by global competition and to draw research, new businesses and private investment to the state. In work done last year, my colleague and I estimated, using conservative assumptions, that New Jersey could realize significant economic benefits in the form of additional jobs and income as a result of the stem cell investment proposed at the time. In addition, as part of a worldwide effort to develop stem cell therapies, New Jersey could realize further benefits from reduced health care costs, lower rates of premature death and improvements in workforce productivity.

With human life at issue, politics must step aside and explicitly not seek to serve local interests with marginal infusions of scarce resources.

Sustaining Research Dollars

Second, it is important that New Jersey maintain a sustained and significant investment in stem cell research. Gov. Jon Corzine's proposed Edison Innovation Fund represents an excellent way to make strategic investments over time in new science and technology initiatives, including stem cell research. One danger posed by the lack of enhanced federal support is that the stem cell initiatives in various states may fall victim to cyclical state budget constraints and local political jockeying on how and where state resources are spent. Such effects could create damaging start-stop support and interfere with scientific peer review of the research. New Jersey must avoid such pitfalls at all costs. Third, the nature of scientific research requires flexibility and openness. Peer review decisions must guide New Jersey's research so that it can benefit from the serendipitous nature of scientific discovery whereby new directions of science arise from unexpected and unplanned outcomes of basic research. This argues for investments in basic stem cell science without attempting to select "winning" areas of investigation.

Fourth, New Jersey needs to act now, with the single-minded goal of becoming a serious, first-rate, competitive participant in stem cell research. This means that scarce state resources should be concentrated and focused on a single coordinated effort, not diffused and dispersed. With human life at issue, politics must step aside and explicitly not seek to serve local interests with marginal infusions of scarce resources that are highly unlikely to yield nationally or globally competitive research capacities.

Economic and Personal Costs of Diseases

In New Jersey, an estimated 6 million disease or injury years can be expected over the next decade and a half from those conditions most likely to benefit from stem cell therapies: Type 1 and other forms of diabetes, Parkinson's disease, Alzheimer's disease, stroke, heart attack and spinal cord injury. Health impacts at this scale have enormous economic costs. Behind this imposing but impersonal number lie the individual burdens and pain of the afflicted and their families who now are, and others in the future who will be, affected daily by these conditions.

There is probably not a single New Jerseyan who has more than one degree of separation from direct experience with these diseases and injuries suffered by a spouse, a child, a family member, a colleague or a friend. New Jersey has a responsibility and an opportunity to affirm life by investing in research that can hold the key to alleviating such pervasive and deeply painful human suffering.

5

Misinformation About Stem Cell Research Is Slowing Funding and Progress

Alan Boyle

Alan Boyle is an award-winning science writer, the science editor at MSNBC.com, and a board member of the Council for the Advancement of Science Writing.

Although there is hope that embryonic stem cell research will quickly lead to the ability to transplant these cells into humans to cure a variety of illnesses, the reality is that true progress could take years and may never lead to transplantation therapies, says James Thomson, the first scientist to isolate and culture human embryonic stem cells, in an interview with MSNBC.com. Even so, what scientists are learning from stem cell research is vitally important. Thomsom believes that decisions around public support for this research are often based on misinformation, and this is slowing down the progress that scientists can make. New announcements in the media about the possibility of using adult stem cells instead of embryonic stem cells further confuse the public because the potential of adult stem cells will not be realized without first learning more from research on embryonic stem cells.

Seven years ago [1998], when [University of Wisconsin cell biologist] James Thomson became the first scientist to isolate and culture human embryonic stem cells, he knew he was stepping into a whirlwind of controversy.

He just didn't expect the whirlwind to last this long. . . .

In [a June 2005] wide-ranging interview, Thompson explained the reasons behind the research, and touched on many other scientific and moral issues as well. Here is an edited transcript:

MSNBC: How do you see this research developing in the next few years?

James Thomson: I want to make a basic statement first—which almost never gets in the press, but I keep trying—on what I see as the legacy of these cells.

The Scientific Value of Stem Cells

One is the basic science, and simply having better access to the human body. That's the most important legacy. I'm very hopeful that there will be some transplantation applications for this technology, but they're going to be very challenging. And it's been so hyped in the press that people expect it to come the day after tomorrow.

Most of the people who oppose this research, and most of the people who support this research, do it with a profound amount of misinformation.

Ten or 20 years from now, I'm actually currently optimistic that there will be transplantation-based therapies, but even if there was none, and it was a complete failure, this technology is extraordinarily important. . . .

If you think about the kind of transplantation that people talk about—say, Parkinson's—it's a very complex disease. The brain is a complicated organ. It may be possible someday to transplant cells and treat that disease. But it's clearly going to be challenging. When you think about it, that's a pretty crude thing to do anyway. What you want to do is understand how the disease happens in the first place, and you need to prevent or slow its progression to the point where it's not going to be relevant anymore.

Up until now we've never had dopaminergic [dopamine-producing] neurons from humans that we could study in the laboratory. They simply didn't exist. So human embryonic stem cells already give rise to these dopaminergic neurons with very high frequency. I hope that they might be useful for transplantation. They may not be. I mean, you have to be realistic about it. But in terms of a basic model to understand what's happening in the disease, it's unparalleled. I think what will happen is that people will understand the basic biology of the disease using this model system, and they'll come up with therapies, and you won't even know stem cells were involved in creating those therapies. . . .

Lack of Public Understanding

MSNBC: Does it concern you that there are people who say, "We're this close to solving this sort of disease with stem cells, so let's pass this legislation"?

Yeah, it's unfortunate. There are clearly exceptions on both sides, but most of the people who oppose this research, and most of the people who support this research, do it with a profound amount of misinformation. It'd be very nice to clear up that information as much as possible. You can still make an informed choice and be for it or against it, but at least it'd be based on the real facts.

Stem Cells in the Media

When President [George W.] Bush was elected, there were about eight months when stem cells were on the cover of every major newspaper in the United States, because it was prior to 9/11 and it was a slow news period, basically. Nonetheless, when people did polls right at the end of that period, the average American really didn't understand embryonic stem cells, despite that tremendous amount of press coverage, because it just became this political question.

I guess the news media aren't really the media to educate. The news media failed in that role. I don't know how to change it, because every time I have an interview with some guy and try to go through what the science is, they talk about curing Alzheimer's.

The Moral Debate

I know you've given a lot of thought to the moral questions and the issues that the opponents of this research bring up. People who have questions about the research would ask, "Is it really worth it to go down this long road where we don't see cures just over the horizon, and we're using components of human life?"

You have to take the other point of view seriously. Nonetheless, the bottom line is that there are 400,000 frozen embryos in the United States, and a large percentage of those are going to be thrown out. Regardless of what you think the moral status of those embryos is, it makes sense to me that it's a better moral decision to use them to help people than just to throw them out. It's a very complex issue, but to me it boils down to that one thing.

With embryonic stem cells, that affects most people, and suddenly society has to deal with it.

If you really explain what's happening—that these frozen embryos are ultimately going to be thrown out—almost everybody except those that have to keep to some kind of party line will say, "What's the problem with this? We should go forward with this."

That's separate from creating embryos. That offends a lot more people, and I can understand why. You're creating something that's a tool, and you're making a tool out of this thing.

I haven't seen polls, but just in my personal conversations, using things that are about to be thrown out offends almost no one, including fairly devout Catholics—whereas actually making something into a tool offends a much larger percentage of the population.

From a public policy point of view, it makes a lot of sense to separate these two issues. Part of what's happening, and the reason why things kind of stalled, is that nuclear transfer and therapeutic cloning was intermixed with trying to make new cell lines from pre-existing embryos. They're very separable. There are some scientific reasons why nuclear transfer and therapeutic cloning might have merit, but most of the value of this technology can be captured simply by making cell lines with existing embryos. There's a diminishing return the more deeply you get into it, and if society is not quite ready for both—well, at least take the one where there's a great advantage now and move on with it. . . .

Embryo Creation

What's your view of the idea that the moral issues may become moot because of the things people might be able to do in the future, with blastomeres, or embryos that are altered so that genetically they cannot get beyond a particular stage?

I've talked to the guy from Stanford [University, California, William Hurlbut] who has been a big proponent of that. The basic problem is that we haven't really thought through the implications of Dolly [a sheep, the first mammal ever cloned]. It will take a while to do that. It was 1978 when the first IVF [in-vitro fertilization] child was born, and basically all these issues have to do with that, they have nothing to do with stem cells. It all goes back to IVF. The problem with IVF is that it affects a relatively small part of the population, so society pushed it to the side and didn't deal with it, even though it was controversial at the time. With embryonic stem cells, that affects most people, and suddenly society has to deal with it. So most of the issues go back to IVF and the creation of embryos, and nothing has changed. The moral debate hasn't really changed. If you read the Warnock Commission [a British bioethics investigation that attempted to define the status of a fetus] report, sometime in the '80s, all those issues were brought up.

So Dolly came along, and that did change the moral argument in ways that people haven't fully appreciated. The bottom line is that any cell in your body has a latent potential to form a human being. If I told you there was a frozen cell in liquid nitrogen, and I thawed it out, and I manipulated it, and I put it back into a woman's uterus and it formed a child, could you infer what the moral status of that cell was? Well, no—because that could be true of any cell in your body now. Granted, you have to do a fair amount of manipulation. But in terms of potential, the potential is there. There's no inherent reason why a liver cell can't form a baby.

If you do things with federal money, you're pretty restricted in what you can do.

So arguments based on potential alone are kind of suspect now because of that. It's becoming increasingly difficult to say, "OK, you have this one-cell embryo, how is it fundamentally different from another cell in your body?" If you have a one-cell embryo in a freezer, you have to intervene to actually make it become a child. You have to do a lot to it. You have to thaw it out under very technical conditions and you have to prepare the mother. There are a lot of interventions there. So I think the fundamental problem is that people haven't come to terms with that. This idea that you'll pre-engineer it to fail seems disingenuous to me. At what point is it failing, and just because it's failing, why is that not an embryo? I don't know.

Competition for Research

Then there's the issue of interstate and international competitiveness: Sometimes you get the feeling that there's a race to capitalize on this technology, and then you have the countervailing moral arguments. It's easy to get into this situation where you feel like "we're going to have one hand tied behind our back, because the South Koreans or the British have this line . . ." I assume that you view it differently?

Well, yes and no. There is some advantage about people being worried about the rest of the world getting ahead of us in this. I'd still say that most of the decent papers come out of the United States, but that's changing. Britain, for example, has these very liberal policies, but they're implemented in a way that's actually slowed them down, and they don't have a lot of money to do it. So in spite of these liberal policies, it's still not this golden place to do stem cell research. In Asia, Korea and Singapore are making major investments in this area, and China too. California managed to get the money flowing from Proposition 71 [which allocated state money to fund stem cell research]. That'll make a big difference. So competition does ultimately make the world go round on one level or another, and that's not a bad thing, I guess.

U.S. Policy

Does that enter into the argument over what sorts of procedures would be acceptable for pursuing this?

It certainly should. The United States is just odd. Basically, if you do things with federal money, you're pretty restricted in what you can do. If you do things with private money, you can do whatever you please, even things that would offend most people. So it doesn't appear to me that it represents sound public policy.

In terms of stem cell technology, there was an advance in getting beyond the mouse feeder cells [mouse cells used to culture human cells cause concerns about the potential of disease transmission from mice to humans], but there are still some animal components that need to be used to support human embryonic stem cells. Do you think that problem is solvable?

I think there are going to be a couple of things that will ultimately change federal policy, though we might have to wait for three years. One is that my group and quite a few other groups are developing culture conditions that are much better than the original culture conditions. My prediction is, within the next year there will be at least one and probably several publications demonstrating that: completely defined conditions, no animal products, no feeder layers of either human or animal origins.

Research Progress

The other thing, which is already happening, is that more and more clinically relevant lineages are being published. People have published work with dopaminergic neurons, motor neurons, heart cells—quite a list now. That means that although we're not ready to put these cells into patients yet, the clock is ticking.

And the original cell lines, while they can be put into patients if you jump through enough hoops, they're clearly not the safest things there, because no matter how much testing you do, you might have missed something. If you derive stem cells from day one in completely defined conditions, in the appropriate GLP [Good Laboratory Practice] conditions, there's simply going to be a higher level of safety. So if I were a patient—say, 10 years down the line when the therapies come on board—which cell line would I want? I wouldn't want the original ones.

When we derived those original cell lines, I was very consciously making them just for experimental purposes, because I figured other people would derive them under these GLP conditions and I wouldn't have to deal with that. We'd just do the experiments. Little did I know that we'd be stuck with these cell lines.

The reality is that the federal government, the National Institutes of Health, is the funding that drives basic research and research into new therapies in this country.

It hasn't bothered me yet, because we didn't really know how to make them better. But the field is changing now. We're getting very close to where we can make them under completely defined conditions. And as I said, I suspect there will be multiple people publishing on that. So I hope that that will drive the political process. . . .

The Urgency of Stem Cell Research

It's a long-term process to really characterize these lines well. The existing cell lines have been around for seven years, and people are gearing up to want to do clinical trials. They've been arguing with the FDA [Food and Drug Administration, which regulates stem cell-based therapies] all this time over the original cell lines. Well, that means we want the new cell lines now for some therapy that's going to be 10 years out.

A three-year difference could really hurt people now, because the [culture] media [i.e., the substance in which the stem cells are grown] has gotten qualitatively different. Just deriving the same old cell lines with the same problems made no sense to me, but once those conditions are really defined and safe, then it makes sense to go back and derive more. So I hope public policy allows that.

Current public policy in theory allows it, because you are allowed to use private funding, but the reality is that the federal government, the National Institutes of Health, is the funding that drives basic research and research into new therapies in this country. And if you exclude that, then you're basically stuck. . . .

Stem Cells from Other Sources

How do you respond to the claim that we have these other sources of stem cells—adult stem cells or cord blood—and there's no need to turn to embryonic cells?

We don't. The most studied cell in the whole body, in terms of stem cells, is the hematopoietic [bone marrow] stem cell. It can't be grown. So what you do when you do a bone marrow transplant is you take some bone marrow out of you—actually, we do peripheral blood—and we put it in another patient without expanding it. There's a clinical need for that expansion step, but it can't be done right now. And hundreds of labs for 30 years have studied that adult stem cell, and that's the one we know the most about.

If you go to these other ones, most of them are known by indirect methods, and nobody can actually isolate and expand and grow them in useful ways. But we can already make blood in very reasonable quantities from human embryonic cells. . . .

This divide about adult vs. embryonic, it's a political debate, it's not a scientific debate.

So if you think about particular things, you can find a stem cell that might work for that application, but this ability to expand these cells in an unlimited, stable developmental state is essentially unprecedented among stem cells. . . . People can do mesenchymal stem cells [cells that develop into connective tissue or blood vessels] [from bone marrow] pretty well, and neural stem cells kind of well, but neural stem cells is a good example. If you try to make dopaminergic neurons from fetal neural stem cells, you get a burst of that activity, and it goes away. Nobody's been able to sustainably make dopaminergic neurons from an adult stem cell, or a fetal one, period—whereas with embryonic cells you can do it already. Over time that might change. But a lot of good groups have tried very hard and failed, to date.

And again, getting back to the basic science thing: If we study the embryonic stem cells, we learn the basic science. That knowledge is just as likely to be applied to adult stem cells as to the embryonic stem cells. The knowledge goes back and forth. And in the case of the blood, people have failed at growing that cell for three decades. Well, studying that lineage with embryonic stem cells, we might learn the clues to make it growable, and it might be that we still want to use adult stem cells to do that because there are a lot of advantages to that, but the knowledge might come from embryonic stem cells.

This divide about adult vs. embryonic, it's a political debate, it's not a scientific debate. Scientists choose the model for a question that fits the question. . . . So people who have focused on adult stem cells historically are using embryonic stem cells now. It doesn't mean that's what they think will be the therapy, but they see it as a more useful model to understand the questions they want to ask. . . .

6

Stem Cell Research Is Suffering Due to the Lack of Federal Funding

Sarah Webb

Sarah Webb writes about science, health, and technology.

Stem cell research is a difficult field for scientists to navigate, and many new and even established researchers are too discouraged to risk entering the rocky terrain of ethical debates, restrictive legal guidelines, and funding difficulties. Although the general public is supportive of stem cell research, many government agencies are not—making for a difficult political climate in which to pursue research. Private funds are particularly difficult to secure and public funding falls under restrictive National Institutes of Health guidelines. Still, people are hopeful that the political and possibly even ethical obstacles will be overcome in the near future and the field will be able to advance in the United States.

Sharyn Rossi's entry into neuroscience research using human embryonic stem cells (hESCs) required a full year, a bit of luck, some persistence, and a difficult cross-country move. The New Jersey native applied to 20 graduate programs, including some in nearby New York, but she was unable to find a position close to home. Then in early 2005, she became interested in the work being done in Hans Keirstead's lab at the University of California (UC), Irvine.

Rossi was accepted into UC Irvine's graduate program, but Keirstead didn't have space for her in his lab initially. The move involved some sacrifice, and at first she wasn't sure she had made a good decision. "I wasn't very happy in California, not working [in Keirstead's lab]." she says. She did a summer lab rotation in the lab before starting graduate school, then worked in another lab for the fall quarter. But she continued to talk to Keirstead regularly as she waited for a spot to open. Her chance came in January, at the beginning of the winter quarter. "Now that I'm in his lab, I really love it out here," she says.

Now, just a few months later, her research (which is supported by a fellowship from California's bond-issue-supported California Institute for Regenerative Medicine, or CIRM) is progressing well, but she's nervous about her career in a field in which politics has created a difficult funding climate and an uncertain future.

In the United States, presidential mandates and legal hurdles are major obstacles to funding and carrying out research involving hESCs. State governments—especially California's—and a few private foundations are scrambling to fill the gap. Still, a new generation of stem-cell scientists is forced to contend with restrictions that go well beyond what other early-career scientists—and established stem-cell scientists—must face.

Beginners Take Nothing

National Institutes of Health (NIH) guidelines limit the use of federal funds to work done on hESC lines developed before the 9 August 2001 presidential mandate. Those restrictions include the use of NIH infrastructure—incubators, petri dishes, culture media, and even office supplies. Many of the NIH cell lines are difficult to grow and were developed from mouse feeder cells, making them unusable for clinical applications. Despite those restrictions, NIH continues to play a dominant funding role even for trainees and researchers whose ultimate interests lie beyond research governed by NIH regulations. State funding is a boon to researchers in states such as California, and private funding offers other opportunities, but NIH funds still form the core of many stem-cell research programs.

[M]any of these researchers may decide that the challenges are insurmountable and choose another related field.

Whether it comes from NIH or other sources, experts say the money that's available goes mostly to the few centers— such as the Burnham Institute in San Diego, California, Harvard's Stem Cell Institute, the University of Wisconsin, Madison, and the University of Georgia—that already have the resources needed to carry out large-scale stem-cell research—or to a few other labs that are already recognized for their work on stem cells. "If you are a recognized stem-cell lab, then that is where you either get federal funding or [where] companies or foundations funnel their funding. If you are recognized, you probably don't have as much trouble," says Evan Snyder of the Burnham Institute. Because he works at a major center, Snyder has a lot of discretion to fund trainees in his group to do work they are interested in—although a project must, of course, fall within NIH guidelines if NIH money or infrastructure is used.

For researchers just starting out, finding funding is much harder. "Where I think you're going to see a problem is the new lab that wants to get into stem-cell research that hasn't previously been doing it or a lab that may be doing stem-cell research but is in a state that really underfunds it," Snyder says. "They need to compete with NIH funding for that topic as they would for any other topic in this climate of not enough money for research." Without this support, he says, many of these researchers may decide that the challenges are insurmountable and choose another related field.

But restrictions on the use of NIH-funded infrastructure may present an even more serious problem for investigators just starting out in the field. New investigators outside the major stem-cell hubs might be able to get a small grant from a private foundation or a state government, but if the work falls outside the NIH-approved guidelines, she won't be able to use existing freezers, incubators, sequencers, or other equipment funded either wholly or in part by government research grants. To separate the work, institutions such as UC Irvine and the University of Wisconsin are constructing buildings to separate research activities that do not fall within NIH rules from those that do. Many researchers see the duplication as wasteful, but under the current regulations, it's necessary.

Funding from an NIH-sponsored training grant and the recognition of the University of Wisconsin's WiCell as a center for embryonic stem-cell research factored heavily into Tom Keenan's decision to take a postdoctoral position there. "I knew that Wisconsin was the place to go because it had WiCell and the access to the stem-cell lines and the stem-cell bank that was not available elsewhere," he says. "At Wisconsin, I knew that I could be successful."

Hedging Their Bets

Because Virginia is not among the states providing substantial funding for stem-cell research, Raj Rao, an assistant professor at Virginia Commonwealth University in Richmond, is seeking funding from private foundations, even though currently all of his lab's work falls within the NIH guidelines. He hopes the future will bring changes to NIH policy, but in the meantime he thinks it's a good idea to "not be totally dependent on NIH funding." Other researchers share Raj Rao's view: They're optimistic that NIH restrictions will be loosened in the near future, but they see wisdom in diversifying their funding portfolios—even in California, where, despite continuing legal challenges, Governor Arnold Schwarzenegger has already advanced $150 million of the promised $3 billion over 10 years to CIRM. The process of disbursing state money to California investigators is moving forward.

Senior researchers often advise trainees to have a backup plan and not to work exclusively with embryonic stem cell lines.

Almost all U.S. researchers work on adult stem cells or mouse stem cells instead of or in addition to hESCs. Senior researchers often advise trainees to have a backup plan and not to work exclusively with embryonic stem cell lines. In some cases, the alternatives are well suited to the research problem or may serve at least as a useful model—but not always. In "my lab, it was always 50% of the people worked on adult stem cells or embryonic stem cells part of the time," says Mahendra Rao. "Almost everyone did stem-cell work, but we mixed and matched so that you were asking a question and that question could be answered with the mouse cells and human cells, or it could be answered with human embryonic cells and adult stem cells." Many students, he says, refused to work with only embryonic stem cells.

Public Engagement

In addition to the practical issues of day-to-day research, many researchers who work with embryonic stem cells deal with the public more than their colleagues in other fields do. "I'm in a field where I sometimes have to keep justifying what I'm doing because there are a lot of public misconceptions," says Raj Rao. "Technically, it is a distraction, but I feel as if it's part of my civic duty as a scientist to get involved in these kinds of issues," says Snyder. Still, "I'm resentful that it's required."

For some graduate students, concern about the difficult funding climate and about public understanding of embryonic stem-cell research have catalyzed advocacy work.

"Even when we have these debates, people are surprisingly open to looking at possibilities," says Mahendra Rao. "I've never had someone in the audience come up to me and tell me that I'm going to die and go to hell because what you're doing is morally and ethically wrong. I've had people come up and tell me that they personally would not want to do this sort of experiment, but they understand our rationale and why they don't think that what we're doing is criminally wrong," he says. The debate has been surprisingly open-minded across the board, he says, even with religious leaders. The sharpest disagreements about the policies result from conversations with U.S government officials, he says.

For some graduate students, concern about the difficult funding climate and about public understanding of embryonic stem-cell research have catalyzed advocacy work. In 2003, Marion Riggs, now a graduate student in Raj Rao's lab, started the Student Society for Stem Cell Research (SSSCR), which now has more than 100 chapters around the United States. Rossi recently founded an SSSCR chapter at UC Irvine.

Stem-cell funding sources—from a 28 July 2006 article in *Science* (subscription req.)

State funding:

- California: $3 billion over 10 years

- Connecticut: $100 million over 10 years

- Illinois: $15 million via executive order

- Maryland: $15 million this year, as a start

- New Jersey: $5 million this year

- Wisconsin: $5 million to attract companies—also involved in building a $375 million facility at WiCell

Private funding:

- Michael Bloomberg: $100 million to Johns Hopkins University

- Starr Foundation: $50 million to Rockefeller University, Cornell University, and Memorial Sloan-Kettering Cancer Center

- Broad Foundation: $25 million to the University of Southern California

- Ray and Dagmar Dolby: $16 million to UC San Francisco

- Stowers Medical Institute: $10 million to Kevin Eggan and Chad Cowan at Harvard

- Leon D. Black: $10 million to Mt Sinai School of Medicine

- Private donors: almost $40 million to Harvard Stem Cell Institute

NIH funding (estimated FY 2006):

- $609 million for stem cells—all types

- $38 million for hESC research

Standing on the Sidelines

Despite the funding obstacles and the political challenges, researchers encourage trainees to enter the field if they're passionate about the science and they've found an interesting question that they want to address. These scientists are optimistic that the political climate will change with future elections and the evolution of public opinion or that scientific progress will make some of the ethical objections obsolete.

But even with this optimism, many of the cutting-edge innovations in embryonic stem-cell research are coming from projects that are not funded by the U.S. government and that fall outside the NIH guidelines, leaving many federally funded researchers watching from the sidelines—and waiting. In the meantime, says Keenan, stem-cell researchers are doing the best they can with what they have: "But I guess in some sense, everyone is looking for a brighter day, trying to hold on until we have full freedom to do what the science tells us we need to be doing."

7

The Lack of Federal Funding Has Not Impeded Stem Cell Research

Yuval Levin

Yuval Levin is the senior editor of The New Atlantis, *a fellow at the Ethics and Public Policy Center (EPPC), and director of the EPPC's program on Bioethics and American Democracy.*

In an effort to combat President George W. Bush's policy on stem cell research, opponents have put forth the false notion that American researchers are quickly falling behind researchers from countries where there are fewer restrictions. Although these claims were refuted scientifically by a group of German scientists, misinformation still continues to abound in the public. In actuality, American researchers are ahead in the field of stem cell research and will continue to lead the way. Although there are some unique regulations in the United States, these regulations do not slow progress. Rather, they exist to ensure that everyone is protected and that science progresses morally, ethically, and responsibly.

Democratic [congressional] leaders Nancy Pelosi and Harry Reid have made clear that overturning President [George W.] Bush's embryonic-stem-cell-research-funding policy will be high on their agenda when they take the reins of the Congress. So come January [2007], it seems we are in for yet another performance of the great stem-cell drama on Capitol Hill.

Yuval Levin, "Falling Behind: Another Embryonic-Stem-Cell Claim Refuted," *National Review Online*, December 4, 2006. Copyright © 2006 by National Review, Inc., 215 Lexington Avenue, New York, NY 10016. Reproduced by permission.

Stem Cell Facts

Opponents of the president's funding policy have by now repeated their lines in this drama so often that every observer has come to know them by heart: It seems 100 million people are sick (*every third American?*), only embryonic stem cells can help them (*based on what proof?*), and by insisting on withholding taxpayer dollars from newly derived lines of cells, President Bush is preventing progress and cures, and causing American scientists to fall behind their counterparts abroad.

This bizarre morality tale is told and retold *ad nauseam*, and has surely sunk in. But now and then, some fragment of fact breaks through the din and threatens the narrative, and for just a brief moment—before that fact, too, is pushed to the side—it seems like the story might fall apart.

The latest such troublesome truth has to do with what is usually the final piece of the great stem-cell narrative: that American scientists are falling behind foreigners because of the Bush-administration's funding policy. That policy, let us recall, does provide (and for the first time) funding for embryonic stem-cell research, but only for lines of cells that existed before the policy came into effect, not for those created after. That way, taxpayer dollars (more than $100 million so far) can advance the research, but without encouraging the ongoing destruction of human embryos.

Their paper, in plain terms and lucid tones, utterly demolishes the notion that American scientists are the slow runners in the global embryonic-stem-cell race.

Researchers in the United States Are Not Falling Behind

This one ethical limit, say opponents of the policy, sets American scientists behind their foreign counterparts in the embryonic-stem-cell race. "The administration's policies have left our researchers far behind the rest of the world," California Senator Dianne Feinstein claimed on the Senate floor in June [2006]. Another Democrat, Rep. Diana DeGette of Colorado, traveled all the way to Britain that same month to assert that "leadership in this area of research has shifted to the United Kingdom."

There is, of course, a simple way to test these claims. Just count the number of stem-cell publications produced by scientists in different countries. In the October issue of the scientific journal *Stem Cells*, a group of German scientists did just that. Their paper, in plain terms and lucid tones, utterly demolishes the notion that American scientists are the slow runners in the global embryonic-stem-cell race.

The team reviewed all original human-embryonic-stem-cell-research publications from 1998 (when such cells were first derived in humans) to the end of 2005. Fully 40 percent (125) of these publications came from one country: the United States. The rest were divided among 20 other nations, with the next nearest competitor (Israel) claiming only 13 percent (42) of the papers. The British, Congresswoman DeGette notwithstanding, came in third with just 9 percent, or 30 publications. A very lopsided lead for America.

Americans Holding the Lead Despite Claims to the Contrary

And the lead seems to be holding, despite prior reports to the contrary. The last major review of embryonic-stem-cell publications, which covered the period from 1998 to 2004, was undertaken earlier this year by two American researchers, Jason Owen-Smith of the University of Michigan, and Jennifer McCormick of Stanford [University, in California], and published in the April 2006 issue of *Nature Biotechnology*. The two clearly set out to prove the claim that Americans were falling behind, and when their data showed otherwise (like this latest study, they found a sizeable American lead) they sought frantically to spin it. Through a series of comical contortions (including comparing American scientists alone to those of the entire rest of the world combined, rather than those in individual countries) they managed to crunch their numbers to show that America's lead is declining. If you squint just right and look sideways at the numbers, such twisted analysis just might let you hold on to the "falling behind" narrative. And indeed, after showing a sizeable American lead, Owen-Smith and Mc-Cormick, without a hint of irony, wrote: "The United States is falling behind in the international race to make fundamental discoveries in hES [human embryonic stem] cell-related fields."

Disputing the Claims

Unlike the more recent German study, Owen-Smith and Mc-Cormick declined to make their full data public (perhaps fearing it would be used as ammunition by supporters of the Bush policy), so it was hard to tell exactly what contortions they engaged in. But the authors of this latest study figured it out. They note that their data does not agree with the previous study's claim that America's lead is declining, pointing out that even if you just count papers published in 2004 or 2005 alone, Americans still published roughly 40% of all embryonic-stem-cell studies. "These divergent findings," the German group writes, "are probably due to the fact that international collaborations of U.S. groups have been marked as 'collaborative research' by Owen-Smith and McCormick." In other words, the previous study excluded from the American count publications on which even one researcher was from a foreign lab, and so arrived at an artificially low number.

The limits on federal funding of embryonic-stem-cell research exist for ethical reasons, not scientific ones.

This latest paper—which, not surprisingly, has received essentially no press coverage—simply and decisively disproves a critical contention of opponents of the Bush policy. But it is important to be clear about exactly what that means.

Limits on Federal Funding Are Essential

The limits on federal funding of embryonic-stem-cell research exist for ethical reasons, not scientific ones. They exist to make sure the government does not endorse the destruction of human life for research, and thus undermine the American ideal of basic human equality. If upholding that principle meant that no stem-cell research at all could proceed, doing so would be no less (or more) justified than it is now. The fact that the principle can be upheld while still enabling so much research to go forward is not the reason the policy is justified. But it is a reason to hope that science and ethics need not stand in opposition to each other. With the right kinds of careful policies, and the right kinds of innovative scientific techniques, science and ethics can go hand-in-hand.

Opponents of the Bush policy, in insisting it sets American scientists behind their foreign counterparts, implicitly argue that science and ethics cannot go hand-in-hand, and that we are forced to choose between them. We now see they are wrong not only in principle, but also in fact.

8

Federal Money and Oversight Are Bad for Stem Cell Research

Sigrid Fry-Revere

Sigrid Fry-Revere is the former director of bioethics studies at the Cato Institute (a think tank, which promotes public policy based on government noninterference) and has written books and articles on the social impact of medical developments.

Stem cell research will make the most progress if it remains privately funded. Although several states passed legislation to fund stem cell research, little of this money has made it to scientists and researchers. Rather, it often gets tied up in legal challenges and extensive public debate. Even when money does make it to laboratories, researchers are often so limited in what they can use the funds for that they are unable to make any significant progress. Additionally, the government is inefficient at determining what should be funded, which slows the distribution of money. Historically, the most progress made in scientific endeavors has been made with private, rather then public funds. As research continues on stem cells, keys decisions should be left to the investors who have a personal interest in ensuring success, and the researchers who understand the science.

Candidates who promised to advance stem cell research fared well in this month's [November 2006] elections. Incoming House [of Representatives] Speaker Nancy Pelosi has

Sigrid Fry-Revere, "Stem Cell Research: Keep it in Private Hands," *The Record*, November 30, 2006, p. L15. Copyright © 2006 North Jersey Media Group, Inc. Reproduced by permission.

said that stem cell therapy has "the biblical power to cure," and others have touted public funding for stem cell research as key to finding cures for everything from Parkinson's disease and multiple sclerosis to hundreds of rare immune and genetic disorders.

Politicians are right that stem cells hold the promise of incredible medical progress. But we won't achieve the promised results if we insist on public funding of the research.

In fact, government-funded research has proved bureaucratic, expensive, wasteful, fickle and divisive.

State Funding Proves Difficult

Since 2003, five states—New Jersey, Ohio, Illinois, Connecticut and California—have pledged billions of dollars to support stem cell research. But political opposition has prevented all but a few million dollars from reaching researchers. This year, six more states pledged funds but none has paid up.

In 2004, California voters approved Proposition 71, which provided $3 billion in public funds for stem cell research. Two years later, challenges to its constitutionality are still pending. Religious and taxpayer groups lost a first round in court, but their appeals, or any other state attempts to fund stem cell research, are likely to be tied up in court for at least another year.

Lucky for California, several private philanthropic organizations have loaned the state $14 million to start providing grants. This confirms the conclusions of a September 2005 article in the *Journal of the American Medical Association*, which found that when public funding for research lapses, private funders almost always step in to take up the slack, often funding projects at a higher rate than did the government.

Other states have had similar experiences. Funds pledged by New Jersey and Illinois—$5 million and $10 million, respectively—are tied up in legislative limbo because politicians can't agree on what types of stem cell research should be funded and where the money should come from. The Ohio governor earmarked $19.4 million for Case Western University [Cleveland, Ohio] for stem cell research, but the Legislature banned it.

A State Example

Only one state got it right: Missouri. This month [November 2006] voters passed a ballot initiative amending the Missouri Constitution to protect the right to pursue and benefit from any stem cell research or therapies allowed under federal law or available to other Americans. Government funding was not the issue; it was the need to guarantee that research could proceed without political interference. The Stowers Institute for Medical Research was standing by with $2 billion in funding. Once the amendment passed, research began within days.

Such private funders are also more efficient than government at doling out research money. After paying administrative costs and interest on the bonds that will have to be issued, Californians will end up forking over more than $6 billion for less than $2.8 billion for stem cell research.

The false hope that the government is taking care of stem cell research will only inhibit private donors and investors from stepping up to the plate.

Private Funding Proves More Successful

To be sure, private organizations attach strings. Many want access to lab financial records, and some discontinue funding if dissatisfied with how research is conducted. But such accountability is important in research. And when stem cell research is not a political football, less time and money is wasted on campaigns, bureaucracy and litigation. More important, those who object to the research aren't forced to fund it.

The development of in vitro fertilization [IVF] provides a perfect model. For years, researchers lobbied government to fund IVF, but amid Luddite [opponents of techology] cries that "test-tube babies" would lead to societal ruin, funding was denied at every turn. As a result, the federal government didn't spend a cent on the development of IVF. Scientists complained that there would be a brain drain away from the United States to countries where such research was publicly funded, and that the United States would fall behind in the development of new reproductive technologies.

None of these gloomy predictions came true. Today, infertility treatment is a $16 billion-a-year industry in the United States.

Funding Must Remain Private

The stem cell debate evokes deja vu [French for "already seen"]. [California-based] Geron Corp. has already invested twice as much in stem cell research as the federal government, and Harvard University [a private university in Massachusetts] has received more requests for money to fund such research than the National Institutes of Health. Some estimate that the stem cell industry will amount to $10 billion by 2010.

Stem cells hold more promise than any breakthrough since [James] Watson and [Francis] Crick discovered the structure of DNA. But governments should not make promises that they cannot keep due to political opposition and their own inefficiency. The false hope that the government is taking care of stem cell research will only inhibit private donors and investors from stepping up to the plate.

We should legalize stem cell research in all its forms, protecting universities and private organizations from the uncertainties of political whims and leave it at that.

9

The Government Should Fund and Regulate Stem Cell Research

Bill Frist

Bill Frist served as a United States senator from 1995 to 2007 and was the Republican majority leader of the Senate from 2003 to 2007. He is also a physician and heart and lung transplant surgeon.

Despite scientific evidence that life begins at conception and the fact that this early life must be destroyed to obtain embryonic stem cells, embryonic stem cell research must continue with the funding and support of the United States. While progress is being made on adult stem cells, embryonic stem cells are unique and essential to realizing the full potential of stem-cell therapies. Further, understanding embryonic stem cells may eventually help scientists better develop adult stem cells, making progress less dependent upon embryonic cells in the future. While embryonic stem cell research continues to make progress in the United States with private funds, the government must step in to support this research. The danger of science progressing without government support is that it allows science to proceed without government oversight. Ethical research principles must be established for several reasons, including ensuring humans are not cloned and embryos are not created solely for the purpose of research, and that research is only conducted on embryos that would otherwise be discarded. The safest, most ethical, and most productive way for embryonic stem cell research to progress is with federal funding and oversight.

Bill Frist, "Stem Cell Research," *Congressional Record*, July 29, 2005.

Since 2001 when stem cell research first captured our Nation's attention, I have said many times the issue will have to be reviewed on an ongoing basis—and not just because the science holds tremendous promise, or because it is developing with breathtaking speed. Indeed, stem cell research presents the first major moral and ethical challenge to biomedical research in the 21st century.

Answering the Moral Questions

In this age of unprecedented discovery, challenges that arise from the nexus of advancing science and ethical considerations will come with increasing frequency. How can they not? Every day we unlock more of the mysteries of human life and more ways to promote and enhance our health. This compels profound questions—moral questions that we understandably struggle with both as individuals and as a body politic.

How we answer these questions today—and whether, in the end, we get them right—impacts the promise not only of current research, but of future research, as well. It will define us as a civilized and ethical society forever in the eyes of history. We are, after all, laying the foundation of an age in human history that will touch our individual lives far more intimately than the Information Age and even the Industrial Age before it.

Defining Life

Answering fundamental questions about human life is seldom easy. For example, to realize the promise of my own field of heart transplantation and at the same time address moral concerns introduced by new science, we had to ask the question: How do we define "death?" With time, careful thought, and a lot of courage from people who believed in the promise of transplant medicine, but also understood the absolute necessity for a proper ethical framework, we answered that question, allowed the science to advance, and have since saved tens of thousands of lives.

So when I remove the human heart from someone who is brain dead, and I place it in the chest of someone whose heart is failing to give them new life, I do so within an ethical construct that honors dignity of life and respect for the individual.

Embryonic stem cells uniquely hold specific promise for some therapies and potential cures that adult stem cells cannot provide.

Like transplantation, if we can answer the moral and ethical questions about stem cell research, I believe we will have the opportunity to save many lives and make countless other lives more fulfilling. That is why we must get our stem cell policy right—scientifically and ethically. And that is why I stand on the floor of the U.S. Senate today.

Four years ago, I came to this floor and laid out a comprehensive proposal to promote stem cell research within a thorough framework of ethics. I proposed 10 specific interdependent principles. They dealt with all types of stem cell research, including adult and embryonic stem cells.

The Importance of Embryonic Stem Cells

As we know, adult stem cell research is not controversial on ethical grounds—while embryonic stem cell research is. Right now, to derive embryonic stem cells, an embryo—which many, including myself, consider nascent [emerging] human life—must be destroyed. But I also strongly believe—as do countless other scientists, clinicians, and doctors—that embryonic stem cells uniquely hold specific promise for some therapies and potential cures that adult stem cells cannot provide.

I will come back to that later. Right now, though, let me say this: I believe today [July 2005]—as I believed and stated in 2001, prior to the establishment of current policy—that the Federal Government should fund embryonic stem cell research. And as I said 4 years ago, we should Federally fund research only on embryonic stem cells derived from blastocysts [early stage of an embryo] leftover from fertility therapy, which will not be implanted or adopted but instead are otherwise destined by the parents with absolute certainty to be discarded and destroyed.

Only with strict safeguards, public accountability, and complete transparency will we ensure that this new, evolving research unfolds within accepted ethical bounds.

Funding Oversight

Let me read to you my fifth principle as I presented it on this floor 4 years ago: No. 5. Provide funding for embryonic stem cell research only from blastocysts that would otherwise be discarded. We need to allow Federal funding for research using only those embryonic stem cells derived from blastocysts that are left over after in vitro fertilization and would otherwise be discarded.

I made it clear at the time, and do so again today, that such funding should only be provided within a system of comprehensive ethical oversight. Federally funded embryonic research should be allowed only with transparent and fully informed consent of the parents. And that consent should be granted under a careful and thorough Federal regulatory system, which considers both science and ethics. Such a comprehensive ethical system, I believe, is absolutely essential. Only with strict safeguards, public accountability, and complete transparency will we ensure that this new, evolving research unfolds within accepted ethical bounds.

Establishing Research Principles

My comprehensive set of 10 principles, as outlined in 2001 are as follows: (1) ban embryo creation for research; (2) continue funding ban on derivation; (3) ban human cloning; (4) increase adult stem cell research funding; (5) providing funding for embryonic stem cell research only from blastocysts that would otherwise be discarded; (6) require a rigorous informed consent process; (7) limit number of stem cell lines; (8) establish a strong public research oversight system; (9) require ongoing, independent scientific and ethical review; (10) strengthen and harmonize fetal tissue research restrictions.

That is what I said 4 years ago, and that is what I believe today. After all, principles are meant to stand the test of time—even when applied to a field changing as rapidly as stem cell research.

The Promise of Stem Cell Research

I am a physician. My profession is healing. I have devoted my life to attending to the needs of the sick and suffering and to promoting health and well being. For the past several years I have temporarily set aside the profession of medicine to participate in public policy with a continued commitment to heal.

In all forms of stem cell research, I see today, just as I saw in 2001, great promise to heal. Whether it is diabetes, Parkinson's disease, heart disease, Lou Gehrig's disease, or spinal cord injuries, stem cells offer hope for treatment that other lines of research cannot offer.

Embryonic stem cells have specific properties that make them uniquely powerful and deserving of special attention in the realm of medical science. These special properties explain why scientists and physicians feel so strongly about support of embryonic as well as adult stem cell research.

Embryonic Stem Cells Are Unique

Unlike other stem cells, embryonic stem cells are "pluripotent." That means they have the capacity to become any type of tissue in the human body. Moreover, they are capable of renewing themselves and replicating themselves over and over again—indefinitely.

Adult stem cells meet certain medical needs. But embryonic stem cells—because of these unique characteristics—meet other medical needs that simply cannot be met today by adult stem cells. They especially offer hope for treating a range of diseases that require tissue to regenerate or restore function.

The Bush Policy

On August 9, 2001, shortly after I outlined my principles, President [George W.] Bush announced his policy on embryonic stem cell research. His policy was fully consistent with my ten principles, so I strongly supported it. It federally funded embryonic stem cell research for the first time. It did so within an ethical framework. And it showed respect for human life.

We should expand federal funding—and thus NIH oversight—and current guidelines governing stem cell research, carefully and thoughtfully staying within ethical bounds.

But this policy restricted embryonic stem cell funding only to those cell lines that had been derived from embryos before the date of his announcement. In my policy I, too, proposed restricting number of cell lines, but I did not propose a specific cutoff date. Over time, with a limited number of cell lines, would we be able to realize the full promise of embryonic stem cell research?

When the President announced his policy, it was widely believed that 78 embryonic stem cell lines would be available for Federal funding. That has proven not to be the case. Today only 22 lines are eligible. Moreover, those lines unexpectedly after several generations are starting to become less stable and less replicative than initially thought; they are acquiring and losing chromosomes, losing the normal karyotype [chromosome distribution], and potentially losing growth control. They also were grown on mouse feeder cells, which we have learned since, will likely limit their future potential for clinical therapy in humans (e.g., potential of viral contamination).

Modifying the Current Policy

While human embryonic stem cell research is still at a very early stage, the limitations put in place in 2001 will, over time, slow our ability to bring potential new treatments for certain diseases. Therefore, I believe the President's policy should be modified. We should expand federal funding—and thus NIH [National Institutes of Health] oversight—and current guidelines governing stem cell research, carefully and thoughtfully staying within ethical bounds.

With more Federal support and emphasis, these newer methods, though still preliminary today, may offer huge scientific and clinical pay-offs.

During the past several weeks, I have made considerable effort to bring the debate on stem cell research to the Senate floor, in a way that provided colleagues with an opportunity to express their views on this issue and vote on proposals that reflected those views. While we have not yet reached consensus on how to proceed, the Senate will likely consider the Stem Cell Research Enhancement Act, which passed the House in May [2005] by a vote of 238 to 194, at some point [during] this [session of] Congress. This bill would allow Federal funding of embryonic stem cell research for cells derived from human embryos that: (1) are created for the purpose of fertility treatments; (2) are no longer needed by those who received the treatments; (3) would otherwise be discarded and destroyed; (4) are donated for research with the written, informed consent of those who received the fertility treatments, but do not receive financial or other incentives for their donations.

Problems with the Stem Cell Research Enhancement Act

The bill, as written, has significant shortcomings, which I believe must be addressed.

First, it lacks a strong ethical and scientific oversight mechanism. One example we should look to is the Recombinant DNA Advisory Committee—RAC—that oversees DNA research. The RAC was established 25 years ago in response to public concerns about the safety of manipulation of genetic material through recombinant DNA techniques. Compliance with the guidelines—developed and reviewed by this oversight board of scientists, ethicists, and public representatives—is mandatory for investigators receiving NIH funds for research involving recombinant DNA.

Because most embryonic stem cell research today is being performed by the private sector—without NIH Federal funding—there is today a lack of ethical and scientific oversight that routinely accompanies NIH-Federal funded research.

Second, the bill doesn't prohibit financial or other incentives between scientists and fertility clinics. Could such incentives, in the end, influence the decisions of parents seeking fertility treatments? This bill could seriously undermine the sanctity of the informed consent process.

Third, the bill doesn't specify whether the patients or clinic staff or anyone else has the final say about whether an embryo will be implanted or will be discarded. Obviously, any decision about the destiny of an embryo must clearly and ultimately rest with the parents.

These shortcomings merit a thoughtful and thorough rewrite of the bill. But as insufficient as the bill is, it is fundamentally consistent with the principles I laid out more than four years ago. Thus, with appropriate reservations, I will support the Stem Cell Research Enhancement Act.

Protecting Human Life

I am pro-life. I believe human life begins at conception. It is at this moment that the organism is complete—yes, immature—but complete. An embryo is nascent human life. It is genetically distinct. And it is biologically human, it is living. This position is consistent with my faith. But, to me, it isn't just a matter of faith. It is a fact of science.

Our development is a continuous process—gradual and chronological. We were all once embryos. The embryo is human life at its earliest stage of development. And accordingly, the human embryo has moral significance and moral worth. It deserves to be treated with the utmost dignity and respect.

I also believe that embryonic stem cell research should be encouraged and supported. But, just as I said in 2001, it should advance in a manner that affords all human life dignity and respect—the same dignity and respect we bring to the table as we work with children and adults to advance the frontiers of medicine and health.

Congress must have the ability to fully exercise its oversight authority on an ongoing basis. And policymakers, I believe, have a responsibility to reexamine stem cell research policy in the future and, if necessary, make adjustments.

Advancements in Science

This is essential, in no small part, because of promising research not even imagined four years ago. Exciting techniques are now emerging that may make it unnecessary to destroy embryos—even those that will be discarded anyway—to obtain cells with the same unique "pluripotential" properties as embryonic stem cells.

For example, an adult stem cell could be "reprogrammed" back to an earlier embryonic stage. This, in particular, may prove to be the best way, both scientifically and ethically, to overcome rejection and other barriers to effective stem cell therapies. To me—and I would hope to every member of this body—that's research worth supporting. Shouldn't we want to discover therapies and cures—given a choice—through the most ethical and moral means?

So let me make it crystal clear: I strongly support newer, alternative means of deriving, creating, and isolating pluripotent stem cells—whether they are true embryonic stem cells or stem cells that have all of the unique properties of embryonic stem cells.

With more Federal support and emphasis, these newer methods, though still preliminary today, may offer huge scientific and clinical pay-offs. And just as important, they may bridge moral and ethical differences among people who now hold very different views on stem cell research because they totally avoid destruction of any human embryos.

Alternative Methods of Obtaining Stem Cells

These alternative methods of potentially deriving pluripotent cells include: (1) extraction from embryos that are no longer living; (2) non-lethal and nonharmful extraction from embryos; (3) extraction from artificially created organisms that are not embryos, but embryo-like; (4) reprogramming adult cells to a pluripotent state through fusion with embryonic cell lines.

Continued Support for Research with Adult Stem Cells

Now, to date, adult stem cell research is the only type of stem cell research that has resulted in proven treatments for human patients. For example, the multi-organ and multi-tissue transplant center that I founded and directed at Vanderbilt University Medical Center [Nashville, Tennessee] performed scores of life-saving bone marrow transplants every year to treat fatal cancers with adult stem cells.

And stem cells taken from [umbilical] cord blood have shown great promise in treating leukemia, myeloproliferative disorders and congenital immune system disorders. Recently, cord blood cells have shown some ability to become neural cells, which could lead to treatments for Parkinson's disease and heart disease.

Thus, we should also strongly support increased funding for adult stem cell research. I am a cosponsor of a bill that will make it much easier for patients to receive cord blood cell treatments.

Embryonic Stem Cells Still Necessary

Adult stem cells are powerful. They have effectively treated many diseases and are theoretically promising for others. But embryonic stem cells—because they can become almost any human tissue ("pluripotent") and renew and replicate themselves infinitely—are uniquely necessary for potentially treating other diseases.

No doubt, the ethical questions over embryonic stem cell research are profound. They are challenging. They merit serious debate. And not just on the Senate floor, but across America—at our dining room tables, in our community centers, on our town squares.

We simply cannot flinch from the need to talk with each other, again and again, as biomedical progress unfolds and breakthroughs are made in the coming years and generations. The promise of the Biomedical Age is too profound for us to fail.

Continuing the Polite Debate

That is why I believe it is only fair, on an issue of such magnitude, that senators be given the respect and courtesy of having their ideas in this arena considered separately and cleanly, instead of in a whirl of amendments and ... maneuvers. I have been working to bring this about for the last few months. I will continue to do so.

And when we are able to bring this to the floor, we will certainly have a serious and thoughtful debate in the Senate. There are many conflicting points of view. And I recognize these differing views more than ever in my service as majority leader: I have had so many individual and private conversations with my colleagues that reflect the diversity and complexity of thought on this issue.

Differing Views

So how do we reconcile these differing views? As individuals, each of us holds views shaped by factors of intellect, of emotion, of spirit. If your daughter has diabetes, if your father has Parkinson's, if your sister has a spinal cord injury, your views will be swayed more powerfully than you can imagine by the hope that cure will be found in those magnificent cells, recently discovered, that today originate only in an embryo.

As a physician, one should give hope—but never false hope. Policymakers, similarly, should not over-promise and give false hope to those suffering from disease. And we must be careful to always stay within clear and comprehensive ethical and moral guidelines—the soul of our civilization and the conscience of our nation demand it.

Cure today may be just a theory, a hope, a dream. But, the promise is powerful enough that I believe this research deserves our increased energy and focus. Embryonic stem cell research must be supported. It is time for a modified policy—the right policy this moment in time.

10

Public Funding Is Problematic for Stem Cell Research

Donna Gerardi Riordan

Donna Gerardi Riordan was the founding director of the Office on Public Understanding of Science at the National Academy of Sciences, and is the current director of programs for the California Council on Science and Technology.

Without federal funding to support stem cell research, many individuals and special interest groups are working at the state level to pass laws that will fund stem cell research and allow it to advance. This process has several unintended consequences, including leaving important stakeholders out of the process, giving too much power to a small group, creating redundancies by requiring that separate facilities be created for work not receiving government funding, causing conflicts with other state interests, and managing the public's high expectations. Additionally, since California passed the Stem Cell Research and Cures Bond Act of 2004, or Proposition 71, very little of the money earmarked for research has actually reached scientists. Rather it is caught up in legal challenges and lawsuits. As other states follow California and try to move forward at the local level, they need to take special caution to avoid the problems and pitfalls experienced in California.

Donna Gerardi Riordan, "Research Funding via Direct Democracy: Is It Good for Science?" *Issues in Science and Technology*, Summer 2008. Copyright © 2008 National Academy of Sciences. Reproduced by permission.

On November 2, 2004, California voters passed the California Stem Cell Research and Cures Bond Act of 2004, popularly known as Prop [Proposition] 71. Its [stated] purpose and intent was to, among other things:

- "Protect and benefit the California budget: . . . by funding scientific and medical research that will significantly reduce state health care costs in the future; and by providing an opportunity for the state to benefit from royalties, patents, and licensing fees that result from the research.

- Benefit the California economy by creating projects, jobs, and therapies that will generate millions of dollars in new tax revenues in our state.

- Advance the biotech industry in California to world leadership, as an economic engine for California's future."

Stem Cell Funding in California

With its passage by 59% of voters, Prop 71 became part of California's constitution. It created the California Institute for Regenerative Medicine (CIRM), a new state agency to administer $3 billion of state bond-funded stem cell research over 10 years. This affront to the [George W.] Bush administration's restrictive policy on embryonic stem cell research received an enthusiastic response from scientists in California and elsewhere in the nation and around the world.

But as is usually the case beyond the initial hype and hope surrounding an emotionally driven public issue, the devil is in the details of designing and implementing good policies. And in this case, much can be learned from the delicate dance in the real world among elected policymakers, advocates for research on specific diseases, public interest groups, and the media. As other states try to emulate California's example, it is important ... to examine the process and the unintended consequences of this approach to funding.

The Consequences of Public Funding

Public funding of a particular research area through the drafting and passage of a topic-specific proposition poses a set of important challenges for research funding, including the unintended consequences of disruption of traditional legislative discussions of thorny public policy issues; fundamental shifts in the involvement and influence of some stakeholders, in this case disease and patient advocates; the Balkanization [division] of research; and managing the expectations of the voters who fund science, particularly if outcomes are not certain. Those interested in science policy, stakeholders, policymakers, and the public ought to consider these challenges as other states consider imitating California's approach.

Other States Follow California's Lead

In states around the nation, research funding decided by direct voter participation is challenging the standard way in which public policy has been made. Traditionally, federal grants flowing from science mission agencies fund most U.S. basic science, primarily via a peer-review system [in which scientists themselves evaluate the research in their fields]. State funding usually supports state universities' research infrastructures or programs of particular importance to the state. Changes in federal and state science funding policies in recent years, however, are causing the creation of new models, such as propositions at the ballot box, that warrant a close examination of their intended and unintended consequences. Whether out of frustration at the slow pace of action, impatience with the intrigue of political gamesmanship, or discomfort with compromises required during the political process, special interest groups and influential individuals are turning to direct democracy, rather than representative democracy, to shape public policy in their own interest.

Other Examples of Direct Democracy

In California, direct democracy occurs primarily through propositions, determined by direct vote of the people at the ballot box. Famous examples of direct voter decisionmaking are Prop 13, which limited the rise of property taxes relative to increases in real estate assessments, and Prop 209, which made racial preferences illegal as a basis for admission to public universities. This direct method of influencing policy emerged nearly 100 years ago during the California Progressive Era [early twentieth century], when individuals, concerned that large businesses such as railroads were having an inordinate influence in the state capital, developed a mechanism to assert some control over policymaking. Successful propositions generally become part of the state constitution, making them exceedingly difficult to amend or repeal.

How California Funded Research

Prop 71 made its way to the ballot primarily through the efforts of Robert Klein, a California real estate developer whose teenaged son is afflicted with juvenile diabetes. Klein was frustrated with the Bush administration's limits on stem cell research and the pace at which the state legislature was moving to fund stem cell research efforts, despite its recently enacted policies intended to draw stem cell research to California. Klein used his considerable wealth and wealthy contacts to underwrite Prop 71's campaign, "Yes on 71: Coalition for Stem Cell Research and Cures." Klein's privately funded effort included drafting the language for Prop 71, securing its inclusion on the fall 2004 ballot through a signature drive, a persuasive multimedia communications effort, flyers mailed to voters, a Web site documenting scores of endorsements from state and national organizations, and strategically timed television ads using celebrities from Hollywood as well as the biomedical research community.

After Prop 71's success at the ballot box, Klein led the effort to establish CIRM at breakneck speed, as specified in the proposition. Prop 71's official language (eight-plus pages of fine print that challenges even those with the best eyesight) stated that the work of CIRM was to be overseen by an Independent Citizens Oversight Committee (ICOC) composed of 29 members representing patient advocate groups, major public and private academic medical centers, private research institutions, and commercial life-science companies in California. Klein was named ICOC's chairman. Prop 71 designated state elected officials to select, within 40 days of the election, ICOC members from specific patient advocacy groups. The governor, for example, selected representatives from Alzheimer's and spinal cord injury advocacy groups, and the president pro tempore [temporary] of the [California] Senate appointed a patient advocate for HIV/AIDS. Other elected officials selected members from specified disease advocacy groups. This potpourri of representation that now governs the workings of CIRM was in and of itself a new experiment in research oversight.

Problems with California's Law

Because of all the attention generated by the Prop 71 campaign, the media and public watchdog groups cast a glaring light on CIRM, a new state agency being established quickly from scratch. CIRM was also emerging under the scrutiny of irked legislators, who, by the proposition's carefully crafted language, were cut out of the process of legislating stem cell research funding, overseeing the formation of CIRM, or revising any of its policies. In fact, Section 8 of Prop 71 prevents the legislature from making any changes until after "the third full calendar year following adoption, by 70 percent of the membership of both houses of the Legislature and signed by the Governor, provided that at least 14 days prior to passage in each house, copies of the bill in final form shall be made available by the clerk of each house to the public and news media." Legislators, displeased by this tying of their hands, showed their dissatisfaction in various ways in the intervening years by trying to impose more oversight and control through a series of legislative attempts, such as regular auditing of CIRM (in addition to the financial reporting required by Prop 71), rules for securing the eggs to be used in research, and the control of intellectual property (IP) derived from state-funded stem cell research.

> *The irony is that in the close to four years after California's grand experiment began with the passage of Prop 71, relatively little new public funding has reached stem cell researchers.*

When the ICOC held its first mandated meeting in mid-December 2004, CIRM existed in name only. It had no fixed location, executive or administrative staff, or guidelines or policies for its work. It also had no funding stream, because the process of selling bonds to fund the agency had not yet been established. The ICOC's first order of business was to hire executives who were both experienced in managing biomedical research and up to the challenges of establishing a new, highly visible state agency. Once hired, that inaugural group's immediate attention turned to the drafting of policies of the three required working groups (scientific and medical research facilities, funding, and medical and scientific accountability standards) for subsequent ICOC approval. Among the issues to be dealt with were conflict of interest, IP rules for recipients of funding from both nonprofit and for-profit institutions, grant submission and peer-review procedures, and a 10-year institutional strategic plan, all of which needed to be drafted, approved, and in place before research grants could be funded. It was no surprise that Klein's initial goal of awarding initial research grants by May 2005, just six months after the November election, was not met.

Funding Delays

In an odd twist of fate, funding was delayed because the state could not sell bonds to fund stem cell research due to a series of legal challenges by anti-stem cell research/antiabortion groups that challenged the constitutionality of CIRM on various grounds. As those cases were consolidated and worked their way through to the state supreme court, CIRM had more time to establish its policies in key areas, and the press, watchdog groups, and legislators had more time to examine every move in detail.

Lawsuits and Challenges

In order to maintain momentum in the face of legal challenges, Klein secured $5 million from private donors to fund CIRM's operations in the first year; during the next year, he raised another $45 million from private donors for what were termed "bond anticipation notes," which would be repaid only if bonds were ever sold. Those funds, which were used for training grants in research institutions around the state, elicited another lawsuit. In July 2006, Governor Schwarzenegger lent CIRM an additional $150 million to begin funding research grants. This move by the governor was politically significant because it occurred one day after President Bush vetoed bipartisan legislation that would have relaxed federal restrictions on stem cell research.

In mid-May 2007, nearly 30 months after its formation, the last legal challenge to CIRM's right to exist and administer $3 billion in research funds ended when the state supreme court refused to hear the appeal of litigation challenging the constitutionality of Prop 71. As a result, the state finally sold the first $250 million in bonds to fund research in October 2007, a full 35 months after the vote that established CIRM.

The irony is that in the close to four years after California's grand experiment began with the passage of Prop 71, relatively little new public funding has reached stem cell researchers in California even though nearly $260 million in grants has been approved for research in 22 institutions around the state. More funding will flow . . . after the approval in May 2008 of $271 million in state funding to build stem cell research facilities in 12 institutions (with $560 million in required matching funding from charitable donations and institutional reserves) and when grants are awarded for new faculty, disease teams, the development of new cell lines, and the creation of new tools and technologies.

Lessons Learned from California

The Prop 71 should serve as a cautionary tale for anyone who believes that more direct democracy always leads to better policy. First, direct voter determination of policy without the input of elected legislators can result in unintended consequences by bypassing the traditional bipartisan and bicameral [Senate vs. House of Representatives] legislative debate about thorny public policy issues.

By excluding legislators from participating in the creation and design of CIRM, the framers of Prop 71 were shortsightedly taunting the state's most powerful and skillful political players. Some legislators, most notably the then-chair of the Health and Human Services Committee, reacted by using their seniority and media savvy to mobilize often rancorous public attention around specific hot-button financial and ethical issues that, in large measure, were already being addressed by CIRM and the ICOC. CIRM, with its newly hired executive staff (who collectively had scores of years of expertise in research administration in universities, industry, and federal agencies) had rapidly begun developing policies and guidelines in all these areas through an open process that was subject to a high level of public scrutiny and accountability through members of the ICOC and the state's open meetings laws. Nevertheless, legislators grabbed the attention of the press and numerous special groups with well-publicized hearings and proposed legislation. Because this effort involved little participation by scientists, it devolved at times into an unfortunate "us versus them" conflict that made it impossible to produce any useful substantive guidance for CIRM and ICOC. A painful irony is that many of the legislators leading efforts to scrutinize CIRM's activities were among the state's earliest and most ardent stem cell research advocates. Given California's strict term limits, these legislators were reluctant to sit on the sidelines of an activity they cared about deeply.

Additional Consequences

A related unintended consequence was the circumvention of critical basic policy processes. In particular, public discussions about IP derived from the bond-funded initiative revealed deep misunderstandings by the legislature, the media, and the public about how basic research is conducted and the ownership of new knowledge derived from creative discovery. For many months, public discussions were bogged down in legislative and media rhetoric asserting that "since we Californians are paying for the research, then we own it." Comments showed up regularly in the media and in hearings that revealed a failure to understand, or an unwillingness to acknowledge, generally accepted federal research grant policies such as the [1980] Bayh-Dole Act that govern IP ownership that provide the nation's current framework for technology transfer resulting in commercial products, or in this case, new treatments and therapies. IP policies for nonprofit and for-profit organizations that will receive grants from CIRM were eventually established, accompanied by vigorous and rancorous public discussions that often produced more heat than light.

Concentration of Power

A second outcome of this exercise in direct democracy is that it concentrated a significant amount of power in a small group of self-designated disease and patient advocates. Those most active in promoting Prop 71 emerged with lasting political power. Whereas a number of disease and patient advocate groups have six- or eight-year terms on the ICOC, there is no formal mechanism for other groups that may have a stake in the advancement of stem cell research to influence policy and direction. Moreover, stakeholders who may have differing opinions, strategies, or priorities are left out of the discussion and have no access other than through highly structured public meetings. The danger of this arrangement is that research priorities are being influenced primarily by their potential applicability to certain disease categories. This is not an effective way to choose basic research projects, and it excludes too many knowlegeable people from the priority setting decisions.

Among states, the creation of special state interests set in policies may impede research or the transfer of research results elsewhere in the nation or around the world.

Problems with New Research Facilities

A third unintended consequence is the Balkanization of research. For example, within research institutions, new facilities will be built with Prop 71 funds. Although these new facilities will usually be welcome at public and private research institutions across the state, they also create a problem. It might be necessary to build in redundancies in facilities to mitigate the risk of sharing facilities that receive federal funding, which does not allow embryonic stem cell research on any cell lines developed after August 2001.[1]

1. On March 9, 2009, President Barack Obama reversed the ban on federally funded embryonic stem cell research in an executive order calling for the National Institutes of Health to create guidelines for funding "responsible, scientifically worthy" human embryonic stem cell research.

Within states, the funds flowing to stem cell research through the proposition process may pit this area of research against other research interests and needs in the state, such as energy, regional climate change, water resources, land use management, and so on. Among states, the creation of special state interests set in policies may impede research or the transfer of research results elsewhere in the nation or around the world. For example, policies that govern the use of IP by firms within California versus outside of California may potentially limit the easy flow of research results across boundaries. And when conflicts or internal inconsistencies exist between state and federal interests and laws, the federal regulations will prevail.

Manipulation of the public to bypass state legislators to support the most recent hot research areas is not a good way to make science policy or public policy.

Public Expectations

The fourth, and arguably the most serious consequence of direct voter participation is the necessity of managing public expectations. The campaign rhetoric in support of Prop 71 created very high expectations. Comments from the public, legislators, public interest groups, and the media after the election exposed unfettered optimism about the pace of creating new knowledge and applying it, reflecting a belief that within 10 years new treatments and therapies would be ready for those with debilitating illnesses. This is an impossibly short time for yielding applications from basic biomedical research. Public comments also revealed the strong expectations that uninsured Californians should have low-cost access to new treatments and therapies developed with any fraction of CIRM funding, and that a percentage of net licensing revenue from them (presumably from a succession of blockbuster products) be returned to the state's general fund. Those expectations are now incorporated in CIRM's policies for IP and revenue-sharing requirements.

The Broader Implications of State-Funded Research

Our system of government was intended to be a representative democracy, an admittedly messy and inefficient system, but one well-designed to put single goals in the context of other state needs. Manipulation of the public to bypass state legislators to support the most recent hot research areas is not a good way to make science policy or public policy. Among the unintended or intended consequences are the lack of deliberation around key issues and a failure to reach consensus about the goals of new public policy.

Moreover, creating a process that excludes legislators and reserves powerful positions for the representatives of specific disease groups unnecessarily limits the inputs and drivers of what ought to be an open, public, and dynamic process that can be responsive to new knowledge and opportunities as they emerge. Stakeholders who were not identified early on will continue to have severely limited ability to make contributions to the policymaking process for stem cell research in California.

When the public becomes disillusioned, it is likely to lose confidence in researchers and research—and not just stem cell research.

Finally, the process of selling a research proposition directly to the public necessarily means that important details such as the time it takes to discover and translate new knowledge into new treatments and therapies and the real costs and benefits of research will not be included in the bumper-sticker sales pitch. We know from decades of research on public understanding of science and technology that although the public doesn't know a lot about scientific or technical issues, in general it trusts the research community. The direct democracy process makes it too easy for scientists to be careless with that trust by becoming participants in campaigns that make unrealistic promises in the quest for research funding.

The Role of Researchers

Although research scientists may not have been the primary drivers of Prop 71, they were willing participants in the campaign to market "stem cell research and cures" to the public. In doing so, they abdicated their responsibility to be sure that the realities of research and innovation were explained to the public.

As for the marketing experts who sold Prop 71 to the public, most have moved on to other nonresearch positions. They will not be held accountable for not meeting the expectations they helped create. When the public becomes disillusioned, it is likely to lose confidence in researchers and research—and not just stem cell research. That is not good for science.

Many in the scientific community regarded Prop 71 as a victory for scientists and research funding. A little reflection will make it clear that making science policy by public referenda is a risky high-stakes game. Basic science research is too important and too dependent on the continuity and stability of public funding to be subjected to the transitory whims reflected in direct democracy.

11

States Must Fund
Stem Cell Research in Lieu
of Federal Support

Liz Barry

Liz Barry is the managing director of the Life Sciences Institute at the University of Michigan.

Although progress in stem cell research is being made in some states, others, like Michigan, have laws that are too restrictive to achieve success. Even though it is likely that there will be increased federal support for stem cell research in the near future, old local state laws may still prohibit some scientists from conducting research in their states or receiving federal funding. These states must act quickly to adopt new laws to enable them to participate in research when it becomes available. Otherwise these states may be left out, which would be detrimental to state economies, inhibit national scientific progress, dissuade researchers hoping to contribute, and would ultimately be harmful to patients waiting for new treatments.

This year, *Time* magazine named University of Wisconsin professor James Thomson one of the 100 most influential people in the world for his work reprogramming adult human cells to take on many of the most promising attributes of human embryonic stem cells. Remarkably, if Professor Thomson had done his work in Michigan rather than Wisconsin he

Liz Barry, "State Stem Cell Policies Deserve National Attention," *Science Progress*, October 15, 2008. www.scienceprogress.org. Copyright © 2008 Center for American Progress. This material was created by the Center for American Progress. www.americanprogress .org.

could have been fined up to $10 million and imprisoned for up to 10 years for his discovery, since the early stages of this work involved the derivation of embryonic stem cell lines and the destruction of human embryos.

Michigan has one of the most restrictive laws in the country with respect to embryonic stem cell research: it is legal for patients to discard human embryos but not legal for scientists to perform research on these discarded embryos even if that is what the patients want. This law delays medical research without saving a single embryo from destruction.

Advocates and opponents of human embryonic stem cell research have heralded Thomson and Shinya Yamanaka's development of the techniques for obtaining induced pluripotent, or iPS, cells. It seems that once the issue of how the cells are derived is off the table, there is widespread agreement that embryonic stem cell research holds great promise for understanding and treating some of our most devastating diseases.

This November, a bipartisan, broad-based coalition in Michigan is trying to change that law with a ballot initiative, known as Proposition 2, that would allow narrowly defined research on human embryos that are leftover after fertility treatment and that would otherwise be discarded if not donated by patients for stem cell research. Although polling suggests that the initiative has strong public support in Michigan, well-financed opponents are pouring millions of dollars into defeating the initiative and keeping the ban in place.

American science succeeds because it is a meritocracy, rewarding achievement and ability over more provincial concerns.

Only a few states have laws as restrictive as Michigan. Both presidential candidates and a majority of members of Congress have affirmed their support for loosened restrictions on embryonic stem cell research, but the Michigan legislature has refused to support the research. Both presidential candidates have recently affirmed their support for embryonic stem cell research, further marginalizing Michigan's policies. The loosening of federal funding restrictions will provide a boost to stem cell research nationwide, but will increase the gulf between scientists in Michigan as compared to those in other states. Unlike scientists at other major research universities, scientists in Michigan universities will remain unable to derive new embryonic stem cell lines for use in expanded federal funding. So is the Michigan vote consequential for national public policy on stem cell research?

Yes.

A peer-governed competitive national system for funding biomedical research has been a fundamental policy and programmatic triumph for the United States. The National Institutes of Health invest over $28 billion each year, 80 percent of which is awarded in peer-reviewed competitive grants to researchers across the nation. This system has advanced our knowledge of disease, led to more effective diagnosis and treatment, and spawned philanthropic and corporate investment, which has fueled our economy. Under this system, the United States has become the global leader in biomedical research. Key to our success has been choosing the best research to fund based on nationwide competition, judged by scientists themselves rather than politicians or lobbyists. American science succeeds because it is a meritocracy, rewarding achievement and ability over more provincial concerns.

This winning approach is thwarted by the patchwork of conflicting state laws and policies regarding human embryonic stem cell research. Forty-five minutes south of Ann Arbor, Michigan, scientists in Toledo, Ohio, are free to use human embryos in research and to derive new stem cell lines. Are the values of Ohio residents so different from the values of Michigan residents (other than in football)? Across the country in California, the state just awarded over $59 million to support the stem cell research of California scientists and has committed a total of $3 billion overall. Massachusetts is investing $1 billion in a similar program, and New Jersey, New York, Wisconsin, Connecticut, and Illinois have also pledged millions of dollars in funding. Yet scientists in Michigan would go to jail for doing the work for which scientists in these states are receiving millions of dollars in state funding.

A comprehensive national system is needed to make the quickest progress in harnessing the potential of human embryonic stem cells to improve the treatment of disease.

Diverse and separate state funding undercuts the successful system of choosing which research to fund based on nationwide competition and peer review. California scientists are only competing with other California scientists for the funds available there and Illinois scientists will only compete with other Illinois scientists. Scientists in other states, who may sometimes have greater expertise, will not have the opportunity to help solve the important problems targeted by these states for funding. This fractured system is antithetical to the goal of funding the most meritorious research and to engaging all of our resources in the war against disease. Families affected by disease don't care where the cure comes from. Yet under the current system, geography drives research investment and determines the problems and approaches that our scientists focus on.

A comprehensive national system is needed to make the quickest progress in harnessing the potential of human embryonic stem cells to improve the treatment of disease. The state-by-state patchwork of funding and regulations is a necessary stop-gap measure to manage the recent lack of federal leadership in this promising area, but ultimately will be self-limiting if not replaced with a more integrated federal approach.

A state-by-state approach to stem cell policy also narrows the opportunities for conducting research that will benefit all segments of our society. One of the more subtle and profound effects of the last seven years of limited federal funding of human embryonic stem cell research is that the stem cell lines currently available for federally funded research are largely derived from embryos obtained from Haifa, Israel. So on one hand, the federal government insists that NIH-funded clinical trials enroll diverse patients that mirror American society, while on the other hand it restricts federally funded scientists to working with embryonic stem cell lines that do not come anywhere close to reflecting the diversity in our society. If embryonic stem cells actually change the future of medicine, we are at risk of leaving some segments of our society out of this future. Who will fix this social justice problem? Scientists in Michigan could, but are prohibited from doing so by state laws. By prohibiting some in our country from working on the important problems we delay progress for all.

> *While we are waiting for federal leadership to prevail, the best we can do is to encourage states like Michigan to bring their policies in line with federal law.*

While we are waiting for federal leadership to prevail, the best we can do is to encourage states like Michigan to bring their policies in line with federal law. If states like Michigan move further away from federal and other state science policies it will be that much more difficult to integrate and engage our scientific community when federal leadership reemerges. Time and talent will be irretrievably lost in the search for new cures.

Michigan is home to the University of Michigan, one of the world's leading research universities. According to various measures of scientific impact, UM is one of the top universities in the world in the field of stem cell research. Yet that impact comes almost entirely from research in the area of adult stem cells. If the upcoming ballot initiative in Michigan fails, it will delay Michigan's ability to develop in the area of pluripotent stem cell research and reinforce the idea that geography should trump merit or promise when the nation determines scientific priorities.

12

Public Funding May Help Reduce the Number of Multiple Births

Liza Mundy

Liza Mundy is a staff writer for The Washington Post *and the author of* Everything Conceivable: How Assisted Reproduction Is Changing Men, Women, and the World.

The process of in vitro fertilization (IVF) and the children created via this method need to be part of the debate over funding for stem cell research. In IVF, a technology that has been in existence for several decades, conception itself takes place in a laboratory and the embryos are implanted into the uterus. Very little is understood, though, about what makes an embryo viable, and the restrictions on embryo research make it unlikely that much will be learned in the near future. This results in doctors needing to implant multiple embryos during IVF to increase the odds of a viable pregnancy, but too often this leads to multiple births. Multiple births not only put the mother at greater risk but also children at risk for preterm delivery, cerebral palsy, and infant mortality. Further, the embryos that are not implanted are the embryos used for embryonic stem cell research, and understanding these embryos is essential for progress in stem cell research. As embryonic stem cell research is debated, the process of creating life through IVF must be part of the debate as well.

The human womb is, ideally, a single-occupancy dwelling. One baby at a time is what women's bodies are marvelously calibrated to conceive and carry. One baby has lots of room for brain growth and organ development; one baby is (relatively) easy to deliver; one baby will usually have at least nine months of close parental bonding before another sibling possibly comes along.

The Rise in Multiple Births

Yet in the past 30 years, this country has experienced a stunning escalation in multiple births. The number of babies born as triplets, quadruplets or even more rose from about 900 in 1972 to 7,275 in 2004. That same year, the highest number of twins ever were born—132,000, nearly double the number born in 1980. Not coincidentally, there has also been a rise in premature births, infants born with low birth weights and disorders—such as cerebral palsy—that can occur when a premature baby's brain is insufficiently developed.

In the debate over federal funding for embryonic stem cell research, these facts need to be talked about. The broad U.S. ban on embryo research funding is one major undiscussed cause of our epidemic of multiple births. We need to consider embryo research as something important to the health of mothers and infants.

One factor among several in the rise in multiple births is in vitro fertilization [IVF], the popularity of which has soared. This is not surprising, since one in seven couples struggles with infertility. There are now 50,000 children born in the United States every year through IVF conception. More than half enter the world as part of a set.

More Information on Embryo Development Is Needed

That's because doctors in the nation's 400 IVF clinics routinely transfer two, three or more embryos into a woman's uterus at a time. They do this because they have no reliable way of telling which embryo has the crucial ability to develop into a fetus.

> *The broad U.S. ban on embryo research funding is one major undiscussed cause of our epidemic of multiple births.*

Since the mid-1970s, when IVF science was getting started, there has been an effective U.S. ban on funding for embryo research. At first a series of presidential commissions debated the ethics while postponing a decision; a board was created to approve research requests but was quickly disbanded. Scientists could apply for funding, but there was no one to apply to.

In 1996 a law known as the Dickey-Wicker Amendment took effect prohibiting funding research involving the creation or destruction of embryos. The provision is regularly passed as part of the Department of Health and Human Services appropriations bill. It has become a conservative touchstone.

Funding Limitations

The upshot is that scientists who receive federal funding—and most good scientists do—cannot use any part of it, even indirectly, to study the embryos that IVF creates so as to learn how to better assess their viability. "There is so much we do not know about the human embryo that we need to," said scientist James Trimarchi. "The truth is, we really don't know anything."

The Problem with Multiple Births

Many patients are happy to embrace the prospect of multiples, welcoming an instant multi-child family. But the stats are harsh: 50 percent of twins and 90 percent of triplets are born prematurely. Now one in eight U.S. infants is born at least three weeks early, according to a 2006 report by the Institute of Medicine. The preterm delivery rate has risen more than 30 percent since 1981. Twins are six times more likely to suffer from cerebral palsy than are singletons, triplets 20 times more. Infant mortality—death in the first year—is substantially higher in multiple births. One study found that at least one disabled child was produced in 7.4 percent of twin pregnancies, 21.6 percent of triplet pregnancies and 50 percent of quadruplet pregnancies.

And don't forget about the mothers. Among women pregnant with twins, the risks of post-partum hemorrhage and infection are doubled, the risk of death quadrupled. A woman pregnant with twins or triplets is more susceptible to the most dreaded complications of pregnancy, including preeclampsia [a potentially fatal condition involving, among other things, high blood pressure].

Children are born every day whose health and well-being are permanently affected by the funding ban for embryo research.

Also, thanks to the ban, little is known about the health of IVF singletons. In the fall of 2005, a major conference was held by the National Institute of Child Health and Human Development to discuss troubling studies that suggested higher rates of complications and even birth defects among IVF children. There, U.S. scientists acknowledged that there is much they don't know, including whether embryos are affected by the media in which they are cultured, and the long-term impact of the increasingly invasive lab techniques that IVF now often involves.

Embryo Research Beyond Stem Cells

Yet few people are talking about the larger issue of embryo research. The debate is only about stem cells. Congress is expected to hand the White House a bill this week [June 2007] authorizing funding for embryonic stem cell research using excess IVF embryos, with the hope that research on these three-day-old fertilized eggs might yield cures for afflictions such as diabetes and spinal cord injury. The president [George w. Bush] is expected to veto the bill. Polls show a majority of Americans want stem cell research to go forward, so sooner or later, it will. But isn't it time to rethink the broader ban?

Children are born every day whose health and well-being are permanently affected by the funding ban for embryo research. It is puzzling that no advocate has arisen willing to take it on in the name of public health. In England, state-licensed research on IVF embryos is permitted for 14 days after their creation. This limit seems reasonable and worth emulating. Embryos do deserve special moral status. But so does the other group that lacks a voice in this debate: children who owe their lives—and perhaps their afflictions—to the science that made them.

Organizations to Contact

The editors have compiled the following list of organizations concerned with the issues debated in this book. The descriptions are derived from materials provided by the organizations. All have publications or information available for interested readers. The list was compiled on the date of publication of the present volume; the information provided here may change. Be aware that many organizations take several weeks or longer to respond to inquiries, so allow as much time as possible.

Americans for Cures Foundation
550 S. California Ave., Suite 330, Palo Alto, CA   94306
(650) 619-7771 • fax: (650) 833-0105
e-mail: inform@americansforcures.org
Web site: www.americansforcures.org

The Americans for Cures Foundation advocates for stem cell research by clarifying key issues in research, distributing current information, and encouraging decision-makers to fund research and help turn discoveries into cures. The Web site contains basic stem cell facts, latest news links, commentaries, and links to additional resources.

Bedford Stem Cell Research Foundation (BRF)
PO Box 1028, Bedford, MA   01730
(617) 623-5670 • fax: (617) 623-9447
e-mail: info@bedfordresearch.org
Web site: www.bedfordresearch.org

The Bedford Stem Cell Research Foundation is a biomedical institute conducting research both in its own labs and in cooperation with other labs. The foundation's mission is to conduct stem cell and related research for diseases and conditions which currently have no effective treatment methods or cures available. The Web site includes links to recent news on stem cell research and maintains an online library of historically significant articles. BRF also publishes a quarterly newsletter.

California Institute for Regenerative Medicine (CIRM)
210 King St., San Francisco, CA   94107
(415) 396-9100 • fax: (415) 396-9141
e-mail: info@cirm.ca.gov
Web site: www.cirm.ca.gov

The mission of CIRM is to support and advance stem cell research and regenerative medicine under the highest ethical and medical standards for the discovery and development of cures, therapies, diagnostics, and research technologies to relieve human suffering from chronic disease and injury. CIRM was established under the California Stem Cell Research and Cures Initiative, a statewide ballot measure, which called for the establishment of a new state agency to make grants and provide loans for stem cell research, research facilities, and other vital research opportunities. The Web site includes information on funding opportunities and press releases with information about advancements in stem cell research in California.

Center for Genetics and Society (CGS)
436 14th St., Suite 700, Oakland, CA   94612
(510) 625-0819 • fax: (510) 625-0874
e-mail: info@geneticsandsociety.org
Web site: www.geneticsandsociety.org

The Center for Genetics and Society is a nonprofit information and public affairs organization working to ensure public accountability and governance in many areas, including stem cell research. CGS organizes briefings and conferences and publishes reports, fact sheets, and a newsletter on its Web site. The Web site also contains general stem cell research information and links to news articles related to stem cell research.

Coalition for the Advancement of Medical Research (CAMR)
2021 K St. NW, Suite 305, Washington, DC   20006
(202) 725-0339
e-mail: CAMResearch@yahoo.com
Web site: www.camradvocacy.org

The Coalition for the Advancement of Medical Research is a bipartisan coalition comprising more than 100 patient organizations, universities, scientific societies, and foundations advocating for the advancement of breakthrough research and technologies in the field of medical and health research. CAMR's advocacy and education outreach focuses on stem cell research, somatic cell nuclear transfer, and related research fields with the goal of developing treatments and cures for individuals with life-threatening illnesses and disorders. The CAMR Web site includes press releases, poll results, and links to recent news on stem cell research.

Do No Harm: The Coalition of Americans for Research Ethics (DNH)
1100 H St. NW, Suite 700, Washington, DC   20005
(202) 347-6840 • fax: (202) 347-6849
Web site: www.stemcellresearch.org

Do No Harm: The Coalition of Americans for Research Ethics is a coalition of researchers, health care professionals, bioethicists, and legal professionals who support the development of medical treatments and therapies that do not require the destruction of human life, including the human embryo. They seek to educate and inform public policy makers and the general public about ethically acceptable and medically promising areas of research and treatment and to support laws prohibiting the federal funding of research that requires the destruction of human life. Its Web site includes publications such as the *Stem Cell Report*, public and congressional testimonies, press releases, fact sheets, commentaries, and links to articles and government documents related to stem cell research.

Genetics Policy Institute (GPI)
11924 Forest Hill Blvd., Suite 22
Wellington, FL   33414-6258
(888) 238-1423 • fax: (561) 791-3889
e-mail: bernard@genpol.org
Web site: www.genpol.org

The Genetics Policy Institute serves as an information exchange for stem cell experts, patient advocates, and public policy workers. In addition to hosting lectures and other educational initiatives, GPI holds the World Stem Cell Summit—an annual conference of researchers, policy makers, legal experts, and others. GPI publishes the annual *World Stem Cell Report*, the biweekly newsletter *Stem Cell Action News*, and posts current news articles on its Web site.

Harvard Stem Cell Institute (HSCI)
42 Church St., Cambridge, MA   02138
(617) 496-4050
e-mail: hsci@harvard.edu
Web site: www.hsci.harvard.edu

The Harvard Stem Cell Institute was founded in 2004 to establish a cooperative community of scientists and practitioners, as well as to develop new ways to fund and support research. The HSCI conducts research and provides grants to stem cell research projects. It publishes *Stem Cell Lines* three times a year and an electronic newsletter reporting on scientific work published by HSCI principal faculty. The institute also publishes *Stem Book*, an open access collection of peer-reviewed chapters covering a range of topics related to stem cell biology.

International Society for Stem Cell Research (ISSCR)
111 Deer Lake Rd., Suite 100, Deerfield, IL   60015
(847) 509-1944 • fax: (847) 480-9282
e-mail: isscr@isscr.org
Web site: www.isscr.org

The International Society for Stem Cell Research is an independent, nonprofit organization established to promote the exchange and dissemination of information and ideas relating to stem cells; encourage research involving stem cells; and promote professional and public education in all areas of stem cell research and application. ISSCR publishes *The Pulse,* an online newsletter.

New York Stem Cell Foundation (NYSCF)

163 Amsterdam Ave., Box 309, New York, NY   10023
(212) 787-4111
e-mail: Info@NYSCF.org
Web site: www.nyscf.org

The New York Stem Cell Foundation is a nonprofit organization whose mission is to support stem cell research that advances cures, and to support scientists engaged in human embryonic stem cell research and somatic cell nuclear transfer. It works through grants, fellowships, symposia, public outreach, and also seeks to establish new collaborative, state-of-the-art research facilities supported entirely with private funds. The Web site contains links to its newsletter, recent news in the field of stem cell research, basic information about stem cells, and links to information for further study.

Stem Cell Institute (SCI)

McGuire Translational Research Facility, 201 6th St. SE
Mail Code 2873, Minneapolis, MN   55455
(612) 626-4916 • fax: (612) 624-2436
Web site: www.stemcell.umn.edu

The objective of the Stem Cell Institute at the University of Minnesota is to expand the knowledge and the potential of stem cells for both human and animal health. SCI hosts a weekly lecture series and publishes the *SCI Journal.* The Web site includes basic information on stem cells and updates on the Institute's research projects.

Stem Cell Technology Foundation

700 Corporate Center Dr., Suite 201, Pomona, CA   91768
(512) 586-1473 • fax: (512) 233-5494
e-mail: jwang@pscclinical.com
Web site: www.stemtech.org

The Stem Cell Technology Foundation is a nonprofit organization formed by a group of scientists committed to providing resources to stem cell researchers, and helping stem cell centers raise funding to support research for a wide variety of human illnesses, conditions, and injuries. The Web site also contains basic educational information about stem cells and links to current issues in the news regarding stem cells.

WiCell Research Institute

University of Wisconsin-Madison, PO Box 7365
Madison, WI   53707-7365
(888) 204-1782 • fax: (608) 263-1064
e-mail: info@wicell.org
Web site: www.wicell.org

Under the direction of James Thomson, the first researcher to isolate human embryonic stem cells, WiCell seeks to expand the study of human embryonic stem cells by generating knowledge; establishing research protocols; and providing cell lines, research tools, and training to scientists worldwide. WiCell also hosts the National Stem Cell Bank, a repository for stem cell lines listed on the National Institutes of Health Stem Cell Registry. The Web site contains basic information about stem cells as well as updates on the institute's research and other current news in the field.

World Stem Cell Foundation (WSCF)

1104 Camino Del Mar, Suite 14, Del Mar, CA   92014
(858) 483-7731 • fax: (949) 266-8504
e-mail: info@worldscf.org
Web site: www.worldstemcellfoundation.org

The World Stem Cell Foundation is a nonprofit organization dedicated to educating patients and other consumers who want current information and answers to their questions about stem cell research, therapies, and clinical trials. WSCF collaborates with partners from public and private organizations that address specific disease and injury conditions through research, clinical trials, and new treatment therapies. The WSCF Web site currently contains information specific to the role stem cell research might play in HIV/AIDS-related treatment.

Bibliography

Books

Janet T. Arnes, ed. *Stem Cell Research: Issues and Bibliography*. New York: Novinka Books, 2006.

Michael Bellomo *The Stem Cell Divide: The Facts, the Fiction, and the Fear Driving the Greatest Scientific, Political, and Religious Debate of our Time*. New York: Amacom, 2006.

Laura Black *The Stem Cell Debate: The Ethics and Science Behind the Research*. Berkeley Heights, NJ: Enslow Publishers, 2006.

Ewa Carrier and Gracy Ledingham *100 Questions & Answers About Bone Marrow and Stem Cell Transplantation*. Boston: Jones and Bartlett, 2004.

Curt Civin *Understanding the Stem Cell Controversy: Miracle Medicine or Mad Science*. Westport, CT: Greenwood Press, 2008.

Calvin A. Fong, ed. *Stem Cell Research Developments*. New York: Nova Biomedical Books, 2007.

Cynthia Fox *Cell of Cells: The Global Race to Capture and Control the Stem Cell*. New York: Norton, 2007.

Leo Furcht — *The Stem Cell Dilemma: Beacons of Hope or Harbingers of Doom?* New York: Arcade Publishing, 2008.

Eve Herold — *Stem Cell Wars: Inside Stories from the Frontlines.* New York: Palgrave Macmillan, 2006.

Alan Marzilli — *Stem Cell Research and Cloning.* New York: Chelsea House, 2007.

Jennie P. Mather, ed. — *Stem Cell Culture.* Burlington, MA: Elsevier Academic, 2008.

Kristen Renwick Monroe et al., eds. — *Fundamentals of the Stem Cell Debate: The Scientific, Religious, Ethical, and Political Issues.* Berkeley: University of California Press, 2008.

Sally Morgan — *From Microscopes to Stem Cell Research: Discovering Regenerative Medicine.* Chicago: Heinemann, 2006.

Joseph Panno — *Stem Cell Research: Medical Applications and Ethical Controversy.* New York: Checkmark Books, 2006.

Ann B. Parsons — *Proteus Effect: Stem Cells and Their Promise in Medicine.* Washington, DC: Joseph Henry Press, 2004.

Ted Peters — *The Stem Cell Debate.* Minneapolis: Fortress Press, 2007.

Michael Ruse and Christopher A. Pynes — *Stem Cell Controversy: Debating the Issues.* New York: Prometheus Books, 2006.

Christopher   *Stem Cell Now: A Brief Introduction*
Thomas Scott   *to the Coming Medical Revolution.*
                   New York: Plume, 2006.

Susan K. Stewart   *Bone Marrow and Blood Stem Cell*
and Jan Sugar   *Transplants: A Guide for Patients.*
                   Highland Park, IL: Blood and
                   Marrow Transplant Information
                   Network, 2002.

Periodicals

Patrick Barry   "Hold the Embryos: Genes Turn Skin
                   into Stem Cells," *Science News*,
                   November 24, 2007.

Sharon Begley   "Reality Check on an Embryonic
                   Debate," *Newsweek*, December 3,
                   2007.

———   "Multiple Choice: What Condition
                   Could Stem Cells Help First?"
                   Newsweek, July 14, 2008.

Julie A. Burger   "Stem Cells Without Embryos:
                   Solving Dilemmas for Human
                   Rights?" *Human Rights*, Fall 2007.

The Economist   "Me Too, Too: Human Embryonic
                   Stem Cells," November 24, 2007.

Misty Edgecomb   "Scientists Tout Stem Cell Studies,"
                   Bangor Daily News, August 15, 2005.

Sigrid Fry-Revere "No Tax Money for Stem Cells," *Los
                   Angeles Times*, November 28, 2006.

Michael Fumento   "No, the Stem Cell Debate Is Not Over," *American Spectator*, April 2008.

Nancy Gibbs   "Wanted: Someone to Play God," *Time*, March 3, 2008.

Bruce Goldman   "Embryonic Stem Cells 2.0," *Nature Reports: Stem Cells*, May 1, 2008. www.nature.com.

Margaret Goodell   "State Demands Strain US Stem Cell Scientists," *Nature Reports: Stem Cells*, January 3, 2008. www.nature.com.

John Kass   "Abortion at Heart of Stem-Cell Debate," *Chicago Tribune*, November 25, 2007.

Celeste Kennel-Shank   "No Time for Retreat: The Difficulty—and Necessity—of Finding a Middle Ground on Stem Cells," *Sojourners Magazine*, December 2007.

———   "Stem Cells and Human Dignity," *Sojourners Magazine*, March 2008.

Robert Langreth and Matthew Herper   "Stem Cells Get Real," *Forbes*, June 16, 2008.

Alan I. Leshner and James A. Thomson   "Standing in the Way of Stem Cell Research," *Washington Post Online*, December 3, 2007. www.washingtonpost.com.

Yuval Levin   "A Middle Ground for Stem Cells," *New York Times*, January 19, 2007.

Elizabeth Lipp   "Regenerative Therapies Gain
                      Momentum," *Genetic Engineering and
                      Biotechnology News*, October, 15,
                      2008.

Kate Lunau   "How to Reboot the Human Body:
                      In Toronto, a Network of Researchers
                      Work with the Building Blocks of
                      Human Life to Do the Impossible,"
                      Maclean's, November 10, 2008.

Robin McKie   "Religion Must Not Block Progress,"
                      New Statesman, May 19, 2008.

Thomas Nairn   "What Is the Debate About Stem
                      Cells?" *U.S. Catholic*, October 2008.

National Review   "Stem-Cell Success," December 17,
                      2007.

National Right to   "Advances in Adult Stem Cells
Life News   Continue," June 2008.

————   "Study Shows Embryonic Stem Cells
                      Rejected by Immune Systems,"
                      September 2008.

William B. Neaves   "When Does a Person Become a
                      Person?" *National Catholic Reporter*,
                      December 26, 2008.

Joe Palca   "States Take the Lead in Funding
                      Stem-Cell Research," *All Things
                      Considered, National Public Radio*,
                      March 30, 2007. www.npr.org.

Alice Park   "The Quest Resumes," *Time*,
                      February 9, 2009.

Anna Stolley Persky   "Stem Cells: The Changing Legal Environment," *Washington Lawyer*, January 2009.

The Recorder   "The Stem Cell Gold Rush," June 3, 2008.

David T. Scadden and Anthony L. Komaroff   "Will Stem Cells Finally Deliver?" *Newsweek*, December 15, 2008.

Science News   "Potent Promise," September 13, 2008.

Nancy Shute   "3 Ways That Stem Cells May Speed New Cures: First Stem Cell Trial in Humans May Be Followed by End on Federal Ban," *U.S. News & World Report*, January 23, 2009.

Lee M. Silver   "Half Human, Half Cow, All Baloney," *Newsweek International*, May 12, 2008.

Joseph R. Sollee   "Recasting the Federal Debate on Stem Cells," *Genetic Engineering and Biotechnology News*, July 2008.

Sarah Webb   "A Patchwork Quilt of Funding," *Nature Reports: Stem Cells*, November 1, 2007. www.nature.com.

Index